Life of Fred®

Pre-Algebra 0 with Physics

Life of Fred®

Pre-Algebra 0 with Physics

Stanley F. Schmidt, Ph.D.

Polka Dot Publishing

ISBN: 978-1-937032-22-7

Printed and bound in the United States of America

Polka Dot Publishing Reno, Nevada

To order copies of books in the Life of Fred series,

visit our website PolkaDotPublishing.com

Questions or comments? Email the author at lifeoffred@yahoo.com

Fifth printing

Life of Fred: Pre-Algebra 0 with Physics was illustrated by the author with additional clip art furnished under license from Nova Development Corporation, which holds the copyright to that art.

for Goodness' sake

or as J.S. Bach—who was
never noted for his plain
English—often expressed it:

Ad Majorem Dei Gloriam

(to the greater glory of God)

If you happen to spot an error that the author, the publisher, and the printer missed, please let us know with an email to: lifeoffred@yahoo.com

As a reward, we'll email back to you a list of all the corrections that readers have reported.

A Note Before We Begin

In the last chapter of this book, Fred is telling the story of how physicists discovered things about the atom. He tells about how Heisenberg announced his uncertainty principle in 1927: *We will never be able to locate an electron exactly.*

We have lost the electron. When my father was delivering pianos back in 1927, he never lost a single piano.

1927

The first quarter century of our lives is the best time to learn new things. Four-year-olds can learn a new language a lot easier than forty-year-olds.

We have lost the electron.

We might lose pianos.

But we don't want to lose this chance to learn.

In the government schools, physics is often taught in the twelfth grade. Too much time is lost waiting that long. This book fills a big gap. The title of this book might have been *Physics after Arithmetic* or *Physics before Algebra.*

It fits nicely right here →

Life of Fred: Fractions
Life of Fred: Decimals and Percents
Life of Fred: Pre-Algebra 0 with Physics ←
Life of Fred: Pre-Algebra 1 with Biology
Life of Fred: Pre-Algebra 2 with Economics

THREE THINGS FOR MAXIMUM SUCCESS

1. There is a *Your Turn to Play* at the end of each chapter/lesson. Please write out the answers on paper before you look at the Complete Solutions offered on the next page. It will help you to learn. Students who do this often do much better on the **Bridges** that come at the end of every six chapters.

2. Get a sheet of paper and head it with *Formulas*. If your handwriting is really good, *Formulas.* If your handwriting isn't so good, Formulas. You will not be graded on your handwriting skills.*

Write down each formula that you encounter in the book and draw a little picture.

3. I can't think of a third thing.

HOW THIS BOOK IS ORGANIZED

Each chapter is a lesson—40 chapters and six **Bridges** to cross. Enjoy your journey with Fred.

This is the last book in the Life of Fred series in which I ask you not to use your calculator. This is your last chance to practice your addition and multiplication tables. In *Life of Fred: Pre-Algebra 1 with Biology*, you can haul out your calculator and use it to your heart's content.

Reading is the fastest way to learn. It is much faster than watching videos or listening to a lecture.
Just ask my daughter, Jill.

---

* That may come later in life. If you become a teacher and your handwriting at the blackboard looks like this, your students might giggle.

Contents

Chapter 1 Friction. 13
math vs. physics
how to move a safe
nature can surprise us
Chapter 2 Proportional. 17
friction is independent of speed
constants of proportionality
exact speed of light
Chapter 3 One Meter. 21
history of the meter
period of a pendulum
square roots
Chapter 4 c. 25
continuing the history of the meter
hard *c* and soft *c*, hard *g* and soft *g*
constant of proportionality for friction, μ
Chapter 5 Finding Mu. 29
solving $1.7 = 4\mu$
distance equals rate times time, $d = rt$
ducks pronounce μ differently than cows
Chapter 6 Super Fred. 33
when in life to start weightlifting
mathematicians use erasers; physicists shove them
why Super Fred can't push a safe down the hallway
The Bridge (with five tries). 37
Chapter 7 Measuring Force. 43
two types of scales
plotting points
experiments with a spring
Chapter 8 Absolute Truth. 48
how many experiments to prove something is true
hunch, conjecture, theory, law
two suns in our solar system
Chapter 9 Four Ways to Stretch. 51
Hooke's law
breaking rubber bands

Chapter 10 Two Kinds of Friction. 55
kinetic friction vs. static friction
how to compute μ_k and μ_s
why we write from left to right
Chapter 11 Energy. 59
nine forms of energy
the Energy Cards game
storing energy
Chapter 12 A Time for Action. 63
why dolls should not drive bulldozers
how to make it easier for students
to see their teacher
The Bridge (with five tries). 67
Chapter 13 Flabbergasted Fred. 75
slope of a line
Kingie shows Fred a new way to find μ_s
a mathematician—one of the safest occupations
Chapter 14 Normal Force. 79
how you know that you have a liver
adding vectors
resolving a vector into two perpendicular vectors
Chapter 15 Math Makes Things Easier. 85
similar triangles
Kingie proves his new way of finding μ_s is right
Chapter 16 Work. 89
what is work for us is not work for physicists
work equals force times distance
the coldest possible temperature
Chapter 17 Transfer of Energy. 93
squishing Fred into the plastic region
Law of Conservation of Energy
the universe is pure energy
perpetual motion machines
Chapter 18 Storing Energy. 97
photosynthesis
Calories, ergs, joules, and electron volts
one tablespoon of butter
The Bridge (with five tries). 101
Chapter 19 Metric System. 109
1 kilogram doesn't equal 2.20466262 pounds
physicists say that no one in Germany weighs 55 kg

Chapter 20 Measuring Mass. 113
one way to measure mass without having
to count the atoms
1 kg is a chunk of metal in France
Chapter 21 A Second Way to Measure Mass. 117
inertia
Ping-Pong balls hurt less than golf balls
why cramming doesn't work
Chapter 22 Pressure. 121
how to check if you have lost some math ability
official definition of pressure
why getting hit with the point of a corkscrew hurts
Chapter 23 Swim Masks.. 125
area of an ellipse
1 square foot equals 144 square inches
exponents
Chapter 24 Density.. 129
swimming in cotton candy
neutron stars
weight of one atom of iron
Cavalieri's Principle
fluid pressure, p = dh
The Bridge (with five tries). 133
Chapter 25 Why Bubbles Float Upward.. 143
special glass in car windshields
buoyancy, b = dv
Chapter 26 temporarily canceled
Chapter 26 Why Things Sink. 151
finding the volume of a desk lamp
1 cubic foot = 12^3 cubic inches
80-pound rubber ducky
Chapter 27 Nose and Brain.. 155
an advantage of a huge nose
your brain floats
Chapter 28 The Long Straw. 159
air has weight
can't use a straw on the moon
why it is nice that ice floats
Chapter 29 Nature Loves Vacuums. 163
knowing by looking
50-foot glass tubes

Chapter 30 Weighing the Atmosphere. 167
2,123 pounds of pressure on every square foot of Kitty's back
popping Fred's head
The Bridge (with five tries). 171
Chapter 31 Water Fountains. 181
learning etiquette
electric circuits
an elevator definition of an electron
Chapter 32 A Small History. 187
bologna, baloney, and Bologna
dead frogs and electric circuits
why we don't put frogs in our flashlights
Chapter 33 Kitty Café. 191
café or cafe
making schematic drawings of electrical circuits
Chapter 34 Volts and Amperes. 197
batteries don't make electrons
how many electrons in an ampere
resistors
Chapter 35 Ohms. 201
why "I" stands for current
why we wear clothes on days ending in *y*.
Chapter 36 Ohm's Law. 205
series circuits
V = IR reminds Fred of virtue
Chapter 37 Parallel Circuits. 209
computing the ohms of resistors in parallel
proof of the formula for resistors in parallel
The Final Bridge (with six tries!). 215
Chapter 38 Lunch. 227
the number of atoms in a piece of bacon
History of the Atom
Chapter 39 From Strange to Stranger. 233
how to speak metaphorically
more atomic history
Chapter 40 Stranger to Unbelievable. 239
up to 1932, when we knew about four particles
the hundreds of particles since 1932
Answers to **The Bridges**. 245
Index. 285

Chapter One
Friction

Fred always thought that early morning was one of the five best times of the day to go jogging. Other best times were the forenoon*, noon, afternoon, and evening.

It was morning. Fred got on his jogging clothes and headed out his office door. The routine was almost always the same. He would head down the hallway past the nine vending machines, down two flights of stairs, and out into the fresh morning air.

Today was different. He couldn't get down the hallway. A giant safe near the vending machines blocked his way.

There was a note attached to the safe that read, "This is as far as I'm going to deliver this safe. This is too heavy to carry. I quit."

This is silly Fred thought. He shouldn't have tried to carry it. He could have slid it down the hallway.

Fred tried pushing the safe. No dice.**

What kept the safe from moving was friction—where the safe rubbed against the floor. If the safe were floating one inch off the floor, Fred could have pushed it down the hall with one hand.

Fred headed back to his office to think things over and do experiments.

---

* The later part of the morning is called the forenoon.

** *No dice* is an idiom. *No dice* means that it doesn't work or it is useless.

An **idiom** is a phrase in which the individual words do not "add up" to what the phrase means.

When you say, "How do you do?" a person looking at the individual words might respond, "How do I do what?"

When you say, "Let's make a date to go fishing," this is silly. All the dates on the calendar have already been established: January 1, January 2, January 3. . . . If you really made a date, it might be something like January 42.

small essay

Math vs. Physics

If this were just a math problem, Fred would have headed back to his office to think things over. He wouldn't have to do any experiments since math doesn't involve "stuff."

Physics deals with stuff—everything from atoms to stars. If you like to play with the dirt under your fingernails or with 500-pound safes sitting in the hallway, then you will need physics.

There is no stuff in mathematics. You will never be walking in a forest and trip over the number three.

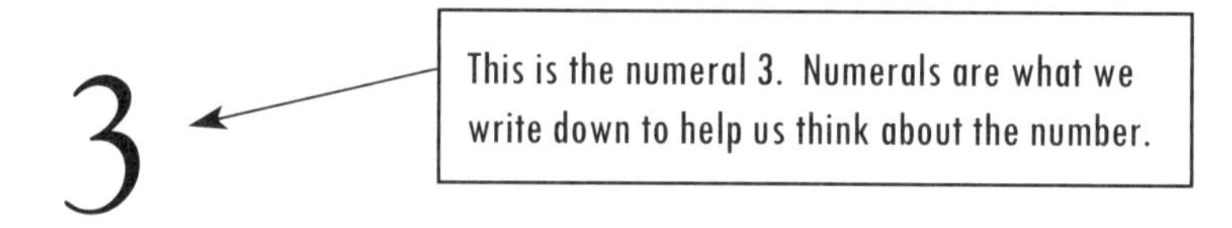

You have never seen a line. Lines are infinitely thin. They are not composed of atoms of ink.

Mathematicians like to say that math is pure.

Physicists might say that math is fairy dust.

Mathematicians will point out that without this fairy dust physicists would be helpless. All of physics needs math.

end of small essay

When Fred got back to his office, Kingie was at work doing another oil painting.

"Your safe has arrived," Fred announced. "It's out in the hallway."

"Wonderful. If you would just bring it in, I would be grateful."

Everyone knows that Kingie wouldn't be much help in moving safes since he is a beanbag doll with no legs.

Fred said, "I'll try and get it moved, but first I need to learn more about friction. This is more than a math problem. The only way to find out about how friction operates is to do experiments in order to find out what is true."

He put three telephone books on his chair so that he would be tall enough to sit at his desk. The desk and chair that KITTENS University had supplied were not designed for teachers who are only three-feet tall.

Fred built a tiny safe that looked just like the real one in the hallway except that it was only two inches (5 cm) tall. He tied a string on it and pulled it across his desktop.

1.9 cm

Then he turned it on its side and pulled it across his desktop.

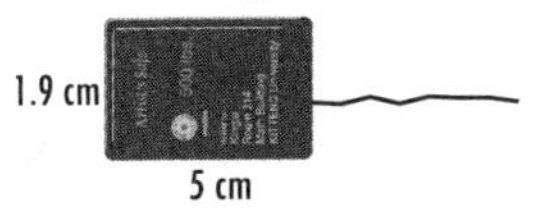

Please write your answers down before you turn the page and see the answers. You will learn a lot more if you do it that way rather than just reading the questions and reading the answers.

Your Turn to Play

1. The dimensions of the tiny safe were 1.9 cm wide, 1.8 cm deep, and 5 cm tall. When the safe was standing upright what is the area of the base that was touching the desktop?
2. When the tiny safe was placed on its side, what was the area touching the desktop?
3. Now I'm going to ask you to guess.

When the safe was lying on its side, the area in contact with the desktop was larger. Did that make it harder to pull the safe?

.......COMPLETE SOLUTIONS.......

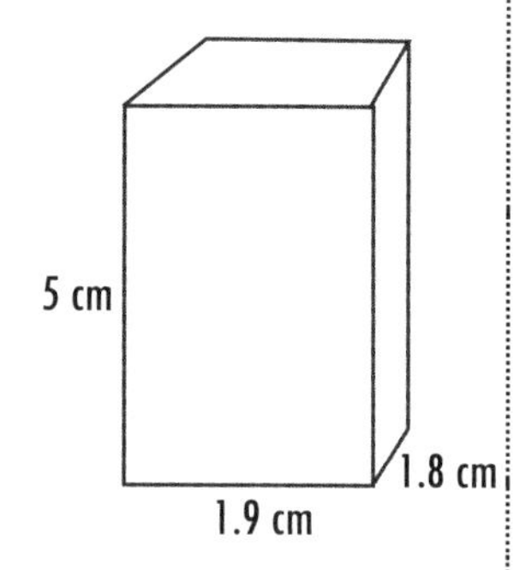

1. The base of the safe is a rectangle. The area of a rectangle is length times width. A = **lw**.

$$
(1.9)(1.8) = \begin{array}{r} 1.9 \\ \times\ 1.8 \\ \hline 152 \\ 19 \\ \hline 342 \end{array} \qquad 3.42 \text{ sq cm}
$$

2. The area of the side of the safe is (5)(1.8).

$$
\begin{array}{r} 1.8 \\ \times\ \ 5 \\ \hline 90 \end{array} \qquad 9.0 \text{ sq cm}
$$

3. A lot of times we won't know what the physical law is until we experiment. Sometimes nature surprises us.

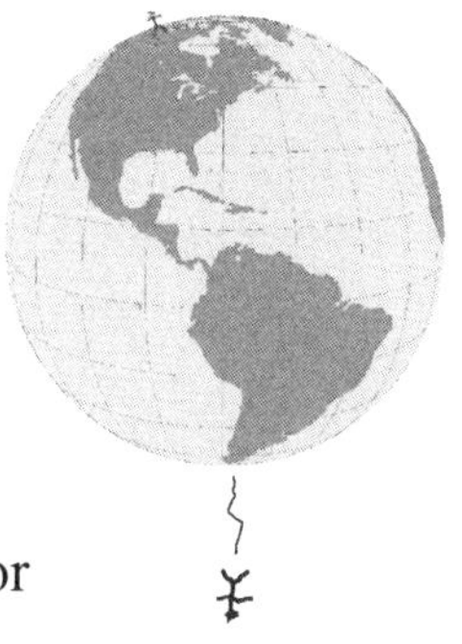

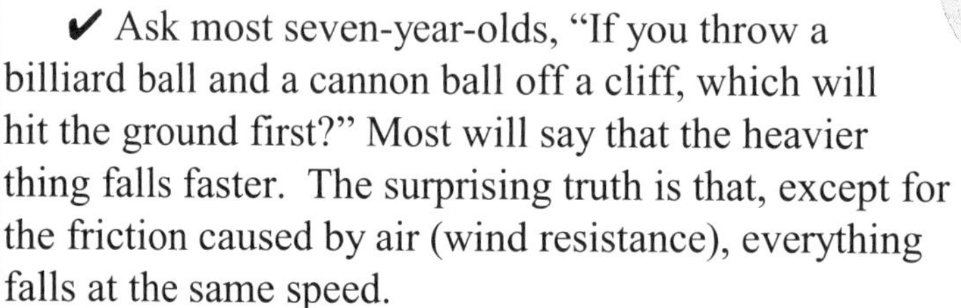

✔ Ask most three-year-olds, "Is the earth flat?" and they will say yes. If you tell them it is a sphere (a ball), they won't believe you and will tell you that people on the bottom would fall off.

✔ Ask most seven-year-olds, "If you throw a billiard ball and a cannon ball off a cliff, which will hit the ground first?" Most will say that the heavier thing falls faster. The surprising truth is that, except for the friction caused by air (wind resistance), everything falls at the same speed.

When Fred did his experiment, he found that it didn't matter whether the safe was upright or on its side. The force he had to apply to the string was the same. Many people would never guess that.

Friction is independent of the area of contact between the two rubbing surfaces.

Chapter Two
Proportional

Fred decided to tip the toy safe on its side. It would not require any more force to push it, but it would be a lot less dangerous for Fred. On its side, the safe would be less likely to fall over and squish him.

Does the speed matter? If the safe is moved quickly, is there less friction than if it is moved slowly? Fred didn't know.

He dragged his toy safe slowly across his desktop.

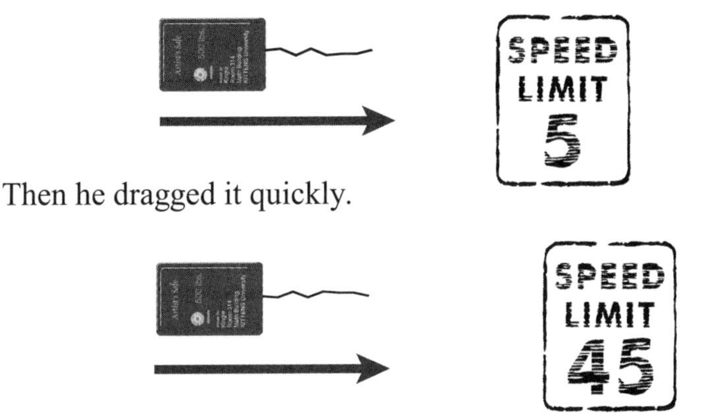

Then he dragged it quickly.

It took a constant pull of 3.22 pounds to keep it moving at the slower speed. It took a constant pull of 3.22 pounds to keep it moving at the faster speed.

Friction is independent of speed. Many people would not have guessed that.

Is friction independent of everything?

Fred built a second tiny safe and stacked it on top of the first one.

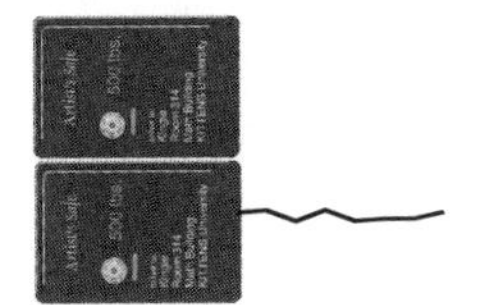

It took a constant pull of 6.44 pounds to keep them moving at a constant speed.

And three safes took 9.66 pounds to keep them moving at a constant speed.

Friction is *not* independent of how hard the safes are pressing down against the desktop. Friction is proportional to the weight. You double the weight and you double the friction. Triple the weight and you triple the friction.

If Kingie's Artist's Safe weighed only 5 pounds instead of 500 pounds, it would be a hundred times easier to push it down the hallway.

There are a zillion things in life (and in physics) in which one thing is proportional to another.

✔ If you are driving at a constant speed, the distance you go is proportional to how long you drive. $D = kT$

(D is distance. T is time. k is called a constant of proportionality.)

✔ If you are paid a fixed amount per month, then the amount you receive is proportional to the number of months you work. $D = kM$

(D is dollars. M is months you worked. k is a constant of proportionality.)

✔ The cost of grapes is proportional to the number of pounds of grapes you buy. $G = kP$

(G is the cost of all the grapes. P is pounds. k is a constant of proportionality.)

✔ If you swim underwater, the pressure on your eardrums is proportional to how deep you are. $P = kD$

(P is the pressure. D is the depth. k is the constant of proportionality.)

✔ If you are pushing a safe down the hallway, the frictional force is proportional to the weight of the safe. $F = \mu N$

Wait! Stop! I, your reader, object. Shouldn't it be $F = kW$***? What is this*** μ ***and why didn't you use*** W ***instead of*** N***?******

I would love to have used k for the constant of proportionality. It would have made a lot more sense. Instead, I've used the Greek letter μ (pronounced mew).

Why? And while we're at it, why don't they use c ***for the constant of proportionality instead of*** k***?***

---

* One of the nice things about this book is that you, the reader, get a chance to talk back to the author. In most physics books, you can scream at the book, but it doesn't do any good.

This book is a little different. I'm listening.

Let's start with your second question. They can't use c for the constant of proportionality. That letter has already been taken. Everybody in physics uses c for the speed of light in a vacuum.

c is exactly 299,792,458 meters per second.

Hold it! This is driving me crazy. It would be fine if you just said that light travels at approximately 299,792,458 meters per second, but you wrote exactly.

Everyone knows that you can't measure things like speed or weight or distance or temperature exactly. You can't tell me my exact height. Those are continuous variables. Things that are reported exactly are described by discrete variables. There are exactly three sheep on this page.

You have given me a lot of things to explain. First of all, in imperial units the speed of light is only approximately 186,282 miles per second.

I'll have to finish up answering these questions in the next chapter because it is now time for . . .

Your Turn to Play

1. Which of these variables are continuous and which are discrete?
   A) the number of sisters you have
   B) the number of English words that you know the meaning of
   C) your shoe size
   D) how many golf balls you can put in your mouth
2. If it takes 5 pounds of force to push two copies of Prof. Eldwood's *Guide to Woodpeckers* across the table at a constant speed if the books are side-by-side, how much force would it take if the books are stacked on top of each other?

5 lbs. ? lbs.

3. If it takes 5 pounds of force to push two copies of *Guide to Woodpeckers* across the table at a constant speed, how much force would it take to push a dozen copies?

.......COMPLETE SOLUTIONS.......

1. If a variable is continuous and can take on values of 7 or 8, then it can take on any value between 7 and 8.

Between any two particular values that a discrete variable can take on, there are values that it can't have. You can have a shoe size of 9 or $9\frac{1}{2}$ but you can't have a shoe size of $9\frac{1}{3}$

All the variables (number of sisters, number of English words, shoe size, and golf balls) are discrete.

2. Friction is independent of the area of contact, so it would take 5 pounds of force to push the books if they were stacked on top of each other.

3. Friction is proportional to weight. A dozen copies would weigh six times as much as two copies, and therefore it would take six times as much force. $6 \times 5 = 30$ pounds of force.

5 lbs.

30 lbs.

I have another small question. In the previous chapter, you wrote that the area of a rectangle is length times width.

Yes, and I wrote it with symbols from algebra: $A = \ell w$.

That's my question. Why did you write part of it in cursive?

If I didn't, it would look like A = lw and that would look like one times w.

I also write b in cursive (ℓ) so it doesn't look like a 6.

Chapter Three
One Meter

Fred took his toy safes and . . . ***Wait a minute! You told me that you would answer my questions at the beginning of this chapter. You forgot your promise.***

I said that I would answer your questions ". . . in the next chapter." I didn't say I would answer them at the beginning of the chapter.

I want my questions answered now!

You mean. . . .

Yes. I mean now.

Okay.

I remember that you were bothered when I said that the speed of light was *exactly* 299,792,458 meters per second. I imagine that you thought that some day with better instruments they would be able to measure c and find that it was 299,792,458.01 meters per second or 299,792,458.00427 meters per second.

That will never happen.

Why?

I'm getting there. A little patience please. This is physics, not math, and things get a bit weird sometimes.

It turns out that physicists *never* measured c and found out it was 299,792,458 meters per second.

That's crazy! How did they find out how fast it was going?

I have to tell a little story. Please don't interrupt.

I won't.

You just did.

Sorry.

Apology accepted.

The Story of the Meter

Back in 1668 (that was 48 years after the Pilgrims landed in Massachusetts in 1620), an English philosopher named John Wilkins (not to be confused with John Wilkes Booth who shot Lincoln) wanted a decimal-based unit of length. He did not especially like the imperial system of 12 inches = 1 foot, 3 feet = 1 yard, 5,280 feet = 1 mile.

His thought was to use the length of a pendulum with a period of two seconds. He called that one meter.

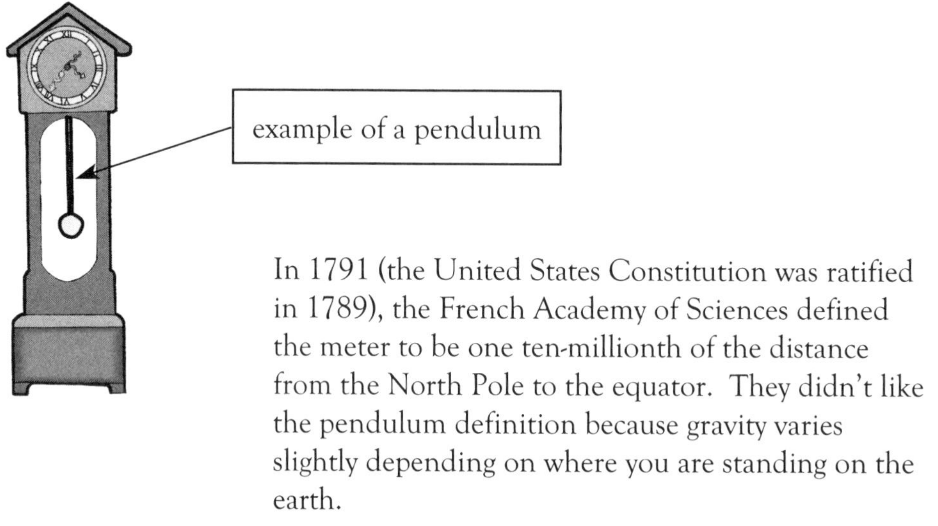

In 1791 (the United States Constitution was ratified in 1789), the French Academy of Sciences defined the meter to be one ten-millionth of the distance from the North Pole to the equator. They didn't like the pendulum definition because gravity varies slightly depending on where you are standing on the earth.

And if gravity varies slightly,
then the period of a pendulum will vary, and
the length of a pendulum with a period of two seconds will vary.

A year later (1792), the French Academy ordered an expedition to measure the earth so that they could determine one ten-millionth ($\frac{1}{10{,}000{,}000}$) of the distance from the North Pole to the equator.

The expedition spent seven years measuring the distance so they could get the length of a meter as accurately as possible.

That would have been nice were the earth a sphere (a ball) or an oblate spheroid (a slightly squashed sphere), but it wasn't. It is a little more irregular than that.

The earth spins.
Here is proof!

As it spins it gets fatter
around the middle.

About a hundred years later (1889), they got tired of the irregular shape of the earth. They made a bar of metal and scratched two lines in it. They said, "That's a meter." And that was the meter until 1960.

I hate to interrupt. Who are "they"?

Some guys in France.*

Then at the Eleventh General Conference on Weights and Measures, which they called *Conférence Générale des Poids et Mesures*, they changed their mind. They wanted something less changeable than two scratches in some metal bar. This was in 1960. They looked at the electromagnetic radiation of krypton. (Krypton is a chemical element like copper, calcium, or helium. Copper has 29 protons in each atom. Calcium has 20 protons. Helium, 2. Krypton has 36 protons. Don't confuse krypton with kryptonite–the stuff that does bad things to Superman. You will never see kryptonite mentioned in any decent science book because it doesn't exist.**)

The Eleventh General Conference declared that one meter is equal to 1,650,763.73 wavelengths of the orange-red emission line in krypton.

Your Turn to Play

1. Make a guess: Does a longer pendulum take more time to swing back and forth than a shorter pendulum?
2. Make a guess: Do two pendulums of the same length but of different weights take the same time to swing back and forth?
3. Is the number of stars you can see at night a continuous variable?

---

* First there was the Meter Convention, which they called *Convention du Mètre* that established the International Bureau of Weights and Measures, which they called *Bureau International des Poids et Mesures*, and then at the First General Conference on Weights and Measures, which they called *Conférence Générale des Poids et Mesures*, they approved the scratched bar as the meter.

That scratched bar is still sitting there at the *Bureau International des Poids et Mesures*, which we call the International Bureau of Weights and Measures.

** Oops! Did I just mention kryptonite?

.......COMPLETE SOLUTIONS.......

1. The **period** of a pendulum is the time it takes for it to make one full trip back and forth. Longer pendulums have longer periods. If we do a bunch of experiments, we might find
(1, 1), (4, 2), (9, 3), and (16, 4).

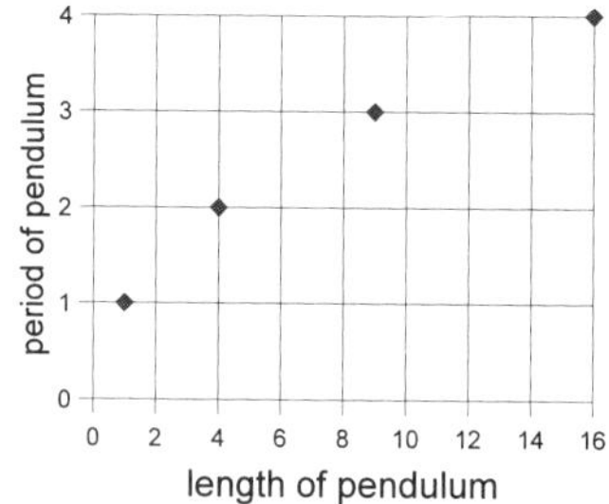

This is not a straight line. The period is not proportional to the length. It is not true that period = k(length).

It is true that period = $k\sqrt{\text{length}}$.*

2. We do a bunch of experiments. We nail a long string to the ceiling of your bedroom. We tie a one-pound weight (such as your sister's shoe) to the end of the string and let it swing back and forth. The period will be three seconds. (Actually, it will be 2.93 seconds if the string is seven feet long.)

Next we tie your sister's two-pound hair dryer and measure the period. Then her three-pound purse. And her four-pound bag of candy. The graph:

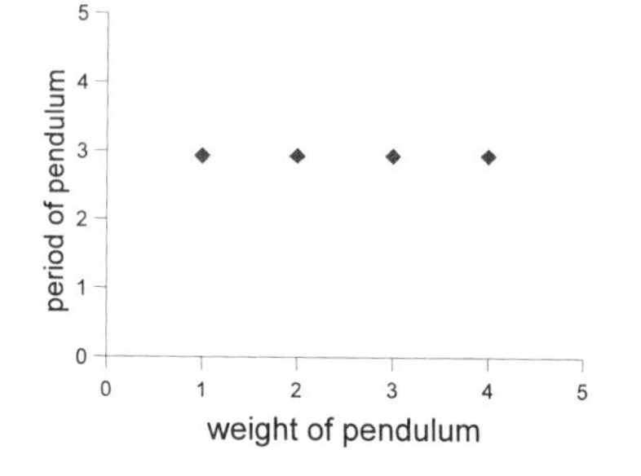

The period is independent of the weight.

3. The number of stars you see will be one of the whole numbers {0, 1, 2, 3, 4, 5, 6, . . .}. You will never see 4.832 stars. You will never see $4\frac{1}{6}$ stars. It is not a continuous variable. It is discrete.

---

* Back in *Life of Fred: Decimals and Percents*, was the first time we encountered the square root symbol $\sqrt{\quad}$.
$\sqrt{4} = 2$. $\sqrt{9} = 3$. $\sqrt{16} = 4$. $\sqrt{49} = 7$. $\sqrt{100} = 10$. $\sqrt{144} = 12$.

Chapter Four

c

Fred took his toy safes and . . . ***Hold it! You spent all of the previous chapter talking about other stuff, and you never answered ANY of the questions I asked in Chapter Two. You were supposed to be working on why light is exactly 299,792,458 meters per second and you just told the story of the meter.***

I was trying to finish up that story, but the Your Turn to Play got in the way.

But what do the Pilgrims landing in Massachusetts in 1620, the 36 protons in Krypton, and the Eleventh General Conference on Weights and Measures in 1960 when they got rid of the metal bar with two scratches have to do with measuring the speed of light exactly?

I'm getting there.

The Story of the Meter
(continued and concluded)

The Seventeenth General Conference on Weights and Measures rolled around in 1983. They were still talking about the length of a meter. They changed the definition again. Since the speed of light in a vacuum is a constant, they declared that a meter is the distance that light can travel in a vacuum in $\frac{1}{299{,}792{,}458}$ of a second. That's a meter. And they have not changed the definition since that meeting.

end of the Story of the Meter

Give me time for this to sink in. Those guys in France . . .

Hold it. They were meeting in France, but this was an international meeting.

Okay. Those guys . . .

I'm assuming you are including both sexes.

Okay. You mean that they weren't measuring the speed of light when they declared that c is exactly 299,792,458 meters per second. They were using the speed of light to declare the length of a meter. And therefore, that velocity is exact. How weird.

Yup.

Now answer the rest of my questions that I asked in Chapter Two.

I've forgotten what you asked.

I haven't. First of all, you just explained why they don't use c for the constant of proportionality, but you never said why they use k.

The answer is that *c* often sounds like *k*.

Ha! You're wrong. The letter c always sounds like k. Candy cane sounds like kandy kane. We could throw away the letter c and just use k. For example: candle, cannibal, canoe, California, car, clam, clap, classic, classroom, clean, cliff, close, cloth, clock, cloud, club, coach, coast, coconut, cocoon, coffeecake, coin, cold, comedy, collect, college.

See, I have proved I'm right. I could give you examples all day long to show you that c is always pronounced like k.

Giving examples all day long doesn't prove anything. I can give you three billion (3,000,000,000) examples of humans that can never become mothers.

You have given examples of hard *c*, which sound like k. In English we also have soft *c*, in which *c* sounds like *s*: census, center, circle, citizen, cycle, cymbal, cell, city, cyan.

You don't happen to know when to use hard c and soft c, do you?

Now that you mention it, I do. The general rule in English is that you have a soft *c* when the *c* comes before *e*, *i*, or *y*.* Of course, because this is English, there are a couple dozen exceptions. This is not like in mathematics where you never divide by zero and two plus two always equals four.

Okay. You have explained why they use k instead of c as the constant of proportionality. But that takes me back to my first question. Since friction is proportional to weight, shouldn't it be $F = kW$ **and not** $F = \mu N$**, which is what you wrote? What is this** μ **and why didn't you use** W **instead of** N**?**

That's two questions, but I won't quibble. I used the Greek letter μ because everyone in physics uses μ as the constant of proportionality for friction. It's just a tradition. If you should ever be forced to look in some

---

*As a bonus we note that there is a hard *g* and a soft *g* in English which follow exactly the same rule—soft *g* before *e*, *i*, or *y*.
Hard *g*: gap, goat, gas, gather, glass, gum, gutter.
Soft *g*: gel, general, giant, gym.

other physics book, you won't be shocked to find μ (pronounced mew) there. Of course, unless it is a really good physics book, you won't find a discussion of hard *c* and soft *c*. And you won't find the Mirror Poem:

I looked in the mirror
And asked,
"Who are you?"
It replied,
"I am μ."

Right now, W (weight) and N (normal force) are exactly the same thing. In a while we will need to change from F = μW to F = μN to keep things true.

Your Turn to Play

1. If μ is equal to 0.25 for steel safes on hallway floors, then how hard would Fred and several of his friends have to push on that 500-pound safe to make it slide down the hallway at a constant speed?

    The formula is . . .

2. If Fred poured oil all over the hallway floor, then μ would equal 0.09. How hard would they have to push?

3. One reason Fred would not pour oil all over the hallway floor is that it would make a mess that he would have to clean up.

    Can you think of a second reason why pouring oil all over the hallway floor, which would reduce μ, would be a bad idea?

4. What is the period as the earth spins?

5. What is the period as the earth goes around the sun?

6. (Harder question) What is the period as the moon goes through all of its phases: new moon ⇨ crescent ⇨ half ⇨ gibbous ⇨ full ⇨ gibbous ⇨ half ⇨ crescent ⇨ new moon?

.......COMPLETE SOLUTIONS.......

1. The formula $F = \mu W$ becomes $F = (0.25)(500)$

$$\begin{array}{r} 500 \\ \times\ 0.25 \\ \hline 2500 \\ 1000 \\ \hline 12500 \end{array} \qquad 125.00$$

The force required would be 125 pounds.

2. $F = \mu W$ becomes $F = (0.09)(500)$

$$\begin{array}{r} 500 \\ \times\ \ 0.09 \\ \hline 4500 \end{array} \qquad 45.00$$

They would only have to apply 45 pounds of force to get the safe moving at a constant speed.

3. The big drawback would be that they could not push very hard. The floor would be slippery, and they would keep falling down.

4. The earth rotates once every day. The period is 24 hours.

5. As the earth goes around the sun, we get the four seasons. The period is one year.

6. From one new moon to the next new moon takes 29.53059 days (29 days, 12 hours, 44 minutes, 2.8 seconds). If you guessed 28 or 29 or 30 days, you get full credit for this question.

μ is pronounced *mew*—a little like saying me-you quickly. There is a second possible pronunciation of μ. You can say it like one of the following animals:

Chapter Five
Finding Mu

Fred took one of his toy safes out into the hallway. He wanted to find out how hard he would have to push on the 500-pound safe in order to move it at a constant speed down the hallway. He wouldn't even have to touch the big safe to find the answer.

He knew that F = μN—that the force of friction equaled the coefficient of friction times the normal force. The normal force is the weight of the safe. In English, μ can also be written as the word *mu*.

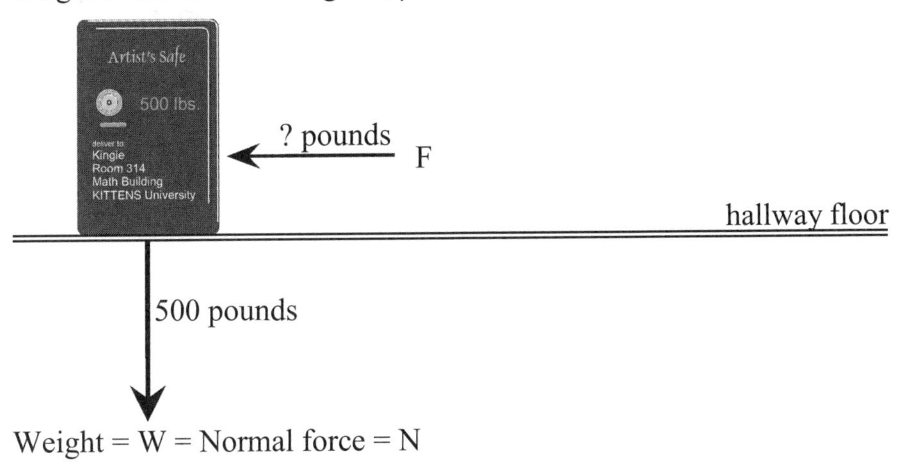

Weight = W = Normal force = N

Fred knew that if μ were equal to 0.25, then F = μN would be F = (0.25)(500) which would be 125 pounds.

Fred knew that if μ were equal to 0.09, then F = μN would be F = (0.09)(500) which would be 45 pounds.

Fred knew that if μ were equal to 0.2, then F = μN would be F = (0.2)(500) which would be 100 pounds.

Guess what Fred didn't know. He didn't know μ. Instead of saying that Fred was clueless, you might say that he was μ-less.

He needed to know mu for a steel safe rubbing against the hallway floor. He could not find that coefficient of friction by just pulling his toy safe across his desktop because steel against a desktop is different than steel against a hallway floor, is different than steel against glass, is different than steel against ice, is different than steel against concrete, is different than steel against steel.

Fred's little toy safe weighed 4 pounds. He pushed it down the hallway at a constant speed. It took 1.7 pounds of force to keep it moving. Sometime I'll show you how to measure that force.

$F = \mu N$ became $1.7 = \mu 4$.

In algebra, we usually write the number in front of the letter. So Fred was looking at $1.7 = 4\mu$.

One way to try and figure out the value of μ that will make $1.7 = 4\mu$ true, is to try various numbers.

If $\mu = 1$, then 4μ would be 4. That is too big.

If $\mu = \frac{1}{2}$ then 4μ would be 2. This is still too big.

If $\mu = 0.3$, then $4\mu = 1.2$. That is too small.

If $\mu = 0.4$, then $4\mu = 1.6$. That is getting closer to 1.7.

An easier way is to divide both sides of $1.7 = 4\mu$ by 4. Since 1.7 and 4μ are equal, if you divide both of them by 4, won't the results have to be equal?

$1.7 = 4\mu$ becomes $\frac{1.7}{4} = \frac{4\mu}{4}$

which becomes $\frac{1.7}{4} = \mu$

Question:
How did I turn $\frac{4\mu}{4}$ into μ?

Answer: If I multiply μ by 4 and then I divide the answer by 4, won't I get back μ? This works with any number. If I take 78935 and multiply it by 4 and then divide the result by 4, I will get back 78935.

And $\frac{1.7}{4} = \mu$

becomes $0.425 = \mu$

$$\begin{array}{r} 0.425 \\ 4\overline{)1.700} \\ \underline{16} \\ 10 \\ \underline{8} \\ 20 \\ \underline{20} \end{array}$$

Let's do that step-by-step:

Start with	$1.7 = 4\mu$
Divide both sides by 4	$\frac{1.7}{4} = \frac{4\mu}{4}$
Simplify	$\frac{1.7}{4} = \mu$
Do the arithmetic	$0.425 = \mu$

Your Turn to Play

1. Using his toy safe, Fred found that the coefficient of friction, μ, for steel against the hallway floor was 0.425. How hard will he need to push against the real 500-pound safe to move it at a constant speed?

2. The distance formula is $d = rt$, where d is the distance, r is the rate, and t is the time. (We saw this formula in *Life of Fred: Decimals and Percents* on page 113. We will use this formula a lot when we get to algebra.)

If a car goes 50 miles per hour (this is its rate) for 3 hours (this is its time), how far will it go?

3. The formula for the coefficient of friction is $F = \mu N$.
F and μN are equal to each other.
If we divide both sides by N, we get $\frac{F}{N} = \frac{\mu N}{N}$
which simplifies to $\frac{F}{N} = \mu$

Now, starting with $d = rt$, do the same thing.
Show the steps to get from $d = rt$ to $\frac{d}{t} = r$.

4. If a car goes 120 miles in 4 hours, what was its rate?

5. Show the steps to get from $d = rt$ to $\frac{d}{r} = t$.

6. Show the steps to get from $F = \mu N$ to $\frac{F}{\mu} = N$.

.......COMPLETE SOLUTIONS.......

1. The formula is $F = \mu N$. We know that μ is 0.425 and N is 500 pounds. $F = (0.425)(500)$. Doing the arithmetic, $F = 212.5$ pounds.

$$\begin{array}{r} 0.425 \\ \times\ 500 \\ \hline 000 \\ 000 \\ 2125 \\ \hline 212500 \end{array}$$

212.500 pounds

Please do not use a calculator in this book.

2. $d = rt = (50 \text{ miles per hour})(3 \text{ hours}) = 150 \text{ miles}$

3. Start with $d = rt$

Divide both sides by t $\frac{d}{t} = \frac{rt}{t}$

Simplify $\frac{d}{t} = r$

4. Using the formula we just derived in the previous problem:

$$r = \frac{d}{t} = \frac{120 \text{ miles}}{4 \text{ hours}} = 30 \text{ miles per hour}$$

5. Start with $d = rt$

Divide both sides by r $\frac{d}{r} = \frac{rt}{r}$

Simplify $\frac{d}{r} = t$

6. Start with $F = \mu N$

Divide both sides by μ $\frac{F}{\mu} = \frac{\mu N}{\mu}$

Simplify $\frac{F}{\mu} = N$

I say mew.

I say μ.

Chapter Six
Super Fred

Using his toy safe, Fred figured out that it would take 212½ pounds of force to move Kingie's new safe down the hallway at a constant speed. Five-year-old, 37-pound Fred would not be able to push that safe down the hallway.

Fred headed back into his office and put his toy safe on his desk. He turned to Kingie who was just finishing up another oil painting. Fred wanted to have a man-to-man conversation with Kingie.*

Fred asked, "Would it be okay with you if we wait a while before I bring in your safe?"

"Sure. I'm in no hurry. This afternoon would be fine."

"I was thinking of waiting a little longer than that," Fred said.

"How long did you have in mind?"

Fred paused and then said, "Could you wait twenty years? Then I will be twenty-five years old and be big and strong."

"Ha!" Kingie exclaimed. "You have been thirty-six inches tall and have weighed about thirty-seven pounds for the last two or three years. When you get to be twenty-five, you might still weigh thirty-seven pounds."

Fred did not want to hear that. Most kids have a mother or a father who can tell them painful truths. Fred had his doll.

Fred went off and sat in a corner of his office. He needed to turn over things in his mind. He thought *Some day I want to be big and tall like Alexander. Maybe I should change my eating habits. Nowadays I hardly eat anything and I drink sugary Sluice. I need more protein for muscles and more calcium for bones.*

*I jog a lot and that's good for heart and lungs. Maybe if I did some weightlifting I would build more muscles.***

---

* Actually, a boy-to-doll conversation would be more accurate.

** Serious weightlifting for kids is not really recommended. It is usually best to wait until teenage years.

Here is what Fred imagined.

Fred at age 5

Fred at age 25 and loaded with muscles

What if I became as strong as Superman? I could lift trucks as easily as a baby lifts a rattle. Pushing a 500-pound safe would be easy.

Fred was wrong. Even if he weighed 250 pounds and had muscles that could push as hard as a locomotive, he wouldn't be able to move Kingie's safe.

That's crazy. I, as your reader, object. If Fred were Super Fred, he could leap over tall buildings, he could run faster than a car, and he could squeeze that safe into a little ball of iron.

I agree. Super Fred could do all of that, but he couldn't push Kingie's safe down the hallway.

I'm listening. Tell me.

First of all, Super Fred wears shoes with rubber soles. That gives him better traction than if he were wearing steel shoes. The coefficient of rubber against the hallway floor is. . . .

Why did you stop?

There is no way to know μ for rubber against the hallway floor except by experimenting. Mathematicians can sit in a comfortable chair with a pencil and a clipboard and invent math. Physicists have to go and shove stuff around.

If you will pardon me for a moment—this eraser is made out of rubber. It weighs 2 ounces. When I shove it along the floor at a constant speed, I have to use 1.6 ounces of force.

$F = \mu N$ is the usual formula. I divide both sides by N and get $\frac{F}{N} = \mu$. I know that N (the weight) is 2, and F (the frictional force) is 1.6.

So $\frac{1.6}{2} = \mu$.

$$\begin{array}{r} 0.8 \\ 2\overline{)\,1.6} \\ \underline{16} \end{array}$$

$\mu = 0.8$ for rubber against the hallway floor

With 250-pound Super Fred wearing shoes with rubber soles, $F = \mu N$ becomes $F = (0.8)(250)$, which is 200 pounds, but it's going to take 212½ pounds to push the safe.

When Super Fred pushes on the safe, he will slide—not the safe.

Your Turn to Play

1. Show that (0.8)(250) does equal 200.

2. Rubber on concrete has a higher coefficient of friction than rubber on a hallway floor.

   Pushing my eraser on concrete, I found that $\mu_{\text{rubber on concrete}} = 0.9$.

   How hard could Super Fred push if he were standing on concrete?

3. If the safe and Super Fred were both out on the concrete sidewalk, he still couldn't push the safe. Why?

4. When physicists study electricity, one of the first laws is V = IR, where V is volts, I is current, and R is resistance. Why don't they write V = cR?

5. If V = IR, then R = ?

6. Which number is larger: 4 or 5 ?

7. Which numeral is larger: 4 or 5 ?

.......COMPLETE SOLUTIONS.......

1. $\begin{array}{r} 250 \\ \times\ 0.8 \\ \hline 2000 \end{array}$ 200.0

2. $F = \mu N$ $F = (0.9)(250) = 225$ pounds. $\begin{array}{r} 250 \\ \times\ 0.9 \\ \hline 2250 \end{array}$ 225.0

3. Super Fred could exert 225 pounds of force against the safe, but the safe is no longer sitting on the hallway floor. It is also on concrete.

We would need to get Fred's toy safe and determine $\mu_{\text{steel on concrete}}$ in order to find out how hard it would be to move the safe.

Since concrete is rougher than a hallway floor, it is probably true that $\mu_{\text{steel on concrete}} > \mu_{\text{steel on hallway floor}}$ (> means greater than.)
So Super Fred could push harder, but the safe would also be harder to push.

4. One reason is that in physics c stands for the speed of light in a vacuum. c is exactly 299,792,458 meters per second.

If they wrote V = cR (instead of V = IR), some people might think that the speed of light was involved.

5. Divide both sides of V = IR by I and you get $\frac{V}{I} = \frac{IR}{I}$

and this simplifies to $\frac{V}{I} = R$

so $R = \frac{V}{I}$

6. 5 is larger than 4. I would rather have 5 pizzas than 4 pizzas.

7. A numeral is the thing you write down when you are thinking about a number.

This is six in Roman numerals: VI
This is six in Arabic numerals: 6
This is six when I'm really tired: 6

4 is a larger numeral than 5.

> Advanced English lesson:
> 7,892,038 is a large number.
> 9 is the largest digit in that number.
> 7,892,038 is one number with 7 digits.
> Writing that number with really small numerals would look like this: 7,892,038

Those of you who have read *Life of Fred: Decimals and Percents* know exactly what this bridge means. You can skip the rest of this page. Just turn to the next page and cross that first bridge.

The rest of you deserve a little explanation.

We are at **The Bridge**.

This is a chance to show that you understand and remember the physics from the beginning of the book up to this point.

You will encounter **The Bridge** after every six or seven chapters.

There are ten questions on **The Bridge**. Answer at least 90% of them correctly, and you can move on to Chapter 7. You are proving that you have mastery (not mystery) of the material. You are permitted to look back to the previous chapters. You just can't look at the answers in the back of the book until you have done all ten problems.

If you don't pass on the first try, then correct your errors on that bridge. That will give you the right to have a second try at crossing the bridge. I've written another ten questions, which I call the *second try*, and you can have another shot at showing you have mastery of the math.

I wrote a third set of ten questions. And a fourth set of ten questions. And a fifth. Pass any one of them, and you have crossed **The Bridge** and can move on to the next chapter.

Some students who really want to learn physics routinely do all **The Bridges** even though they pass on the first or second try.

The Bridge

from Chapters 1–6

first try

Goal: Get 9 or more right and you cross the bridge.

1. Joe is one of Fred's students. He likes to go fishing. When he got to the Great Lake at the edge of the KITTENS campus, he took his rowboat out of his truck and dragged it on the sidewalk. The boat weighed 78 pounds and the coefficient of friction of his wooden boat against the sidewalk was 0.3. How hard did Joe have to pull on the boat to keep it moving at a constant speed?

2. Darlene had come along with Joe on his fishing trip. Joe wanted to show off how strong he was. He asked her to hop into his boat and he would drag both the boat and Darlene. She weighed 110 pounds. How hard would Joe have to pull?

3. When they got to the beach, Darlene hopped out of the boat. Joe could drag the 78-pound boat over the sand with a force of 46.8 pounds. What is the coefficient of friction for the boat against the sand?

4. At the edge of the lake was a sign that the university had put up. It was similar to the sign that some fast-food restaurants have on their coffee cups: "Warning—Coffee is hot."

The sign was 49 cm wide and 37 cm tall. What was its area?

5. Is the number of fish that Joe will catch a discrete variable?

6. Is the weight of the first fish he catches a discrete variable?

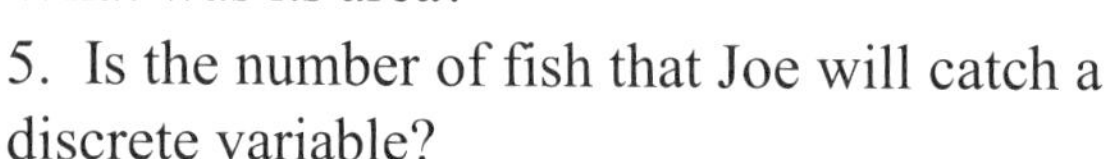

7. gum goes here → Darlene brought her grandfather clock and set it up on the beach. Joe stuck his gum on the pendulum. Will the clock tick slower?

8. How much force is needed to drag a $2\frac{2}{3}$-pound fish on the dock where $\mu_{\text{fish against the dock}} = \frac{1}{3}$?

9. Is 8 a smaller numeral than 5?

10. If $8 = 21\mu$, what does μ equal?

The answers to **The Bridge** are in the back of this book starting on page 245.

The Bridge

from Chapters 1–6

second try

---

1. Darlene dreamed about the day she would marry Joe. There would be a 60-pound wedding cake. She dreamed that it wasn't placed on the middle of the table and Joe shoved it over (leaving a hand print on the side of the cake). The coefficient of friction of a wedding cake on a table is 0.4. How hard did Joe have to push?

2\.

The groom doll fell off the cake as Joe pushed it. The doll weighed 2 pounds. How hard did Joe have to push the cake with the groom doll missing?

3\. In her dream, Joe got scared and she had to drag him up the aisle. He weighed 170 pounds, and she pulled with a force of 34 pounds to keep him moving at a constant speed. What is the coefficient of friction of Joe on the aisle floor?

4\. The wedding bulletin was printed on binder paper in order to save money for Darlene's dress. What is the area of an 8½ x 11-inch sheet of paper?

5\. Is the cost of her dress a continuous variable?

6\. Is the number of bridal attendants a continuous variable?

7\. The distance up the aisle is 50 feet. It took Darlene 8 seconds to drag Joe up the aisle. How fast was Joe moving?

8\. If $14\mu = 9$, what does μ equal?

9\. She dreamed that after the wedding they would head to Disneyland for their honeymoon. How long would they have to drive at 60 miles per hour to go 1600 miles?

10\. If $2.5 = 5x$, what does x equal?

The Bridge

from Chapters 1–6

third try

---

1. One of Fred's favorite poets is Christina Rossetti. She is considered by many to be the best woman poet in England before 1900. Fred pulled the one-pound book of her poetry toward him on his desk with a force of $\frac{1}{3}$ pound. What is the coefficient of friction between the book and the desk?

2. Fred copied some of the lines of Rossetti's "A Royal Princess" on the blackboard in his office.

> ***All my walls are lost in mirrors, whereupon I trace***
> ***Self to right hand, self to left hand, self in every place. . . .***

If Fred wrote at the rate of 3 inches per second, how long could he write in 12 seconds?

3. The cover of the book was very shiny. Fred ran his finger across it pressing with 6 ounces of pressure. If $\mu = 0.19$, how much frictional force did Fred experience?

4. The cover of the book was $4\frac{1}{8} \times 6\frac{3}{4}$ inches. What was the area of the cover?

5. Is the number of Christina Rossetti books that Fred owns a discrete variable?

6. Is the force with which Fred pressed his finger against the cover a discrete variable?

7. If $77\mu = 100$, what does μ equal?

8. Which is a larger number: 8 or 77?

9. If 8 = 43x, what does x equal?

10. Fred hung a 4-foot red thread from the ceiling and tied a 4-pound weight to it. He hung a 5-foot blue thread from the ceiling with a 3-pound weight on it. He gave them each a push. Which has the longer period?

English: Today I hang a picture. Yesterday I hung a picture.
Today I hang a prisoner. Yesterday I hanged him.

The Bridge

from Chapters 1–6

fourth try

1. Jake, Jack, and Jane were three sailors who were visiting New York City. Jake's duffle bag became so heavy that he couldn't carry it. He just dragged it by pulling on the strap.

"What's in your bag?" Jane asked.

"I bought a bunch of bowling balls to give as gifts to my friends," Jake answered.

The bag weighed 150 pounds and Jake could drag it using a force of 50 pounds. What was the coefficient of friction between the bag and the sidewalk?

2. Jake took a 16-pound pink bowling ball out of his bag and gave it to Jane. How much force did Jake now need to drag his duffle bag?

3. Are the number of bowling balls a continuous variable?

4. Is Jack's weight a continuous variable?

5. In 6 hours the three sailors walked 15 miles on the streets of New York. How fast were they walking?

6. Jack really liked electronics. There was one block in the city that just had stores selling televisions, cell phones, computers, etc. The block was in the shape of a rectangle. The four sides of the block were 150 yards, 130 yards, 150 yards, and 130 yards. What was the area of that block?

7. When they got to Central Park, Jack took a rope out of his duffle bag. He climbed up a tall tree and tied one end to a large branch. The rope was 63 feet long. Jake (155 pounds) grabbed the other end of the rope and swung back and forth. One complete round trip (one period) took 7 seconds. After Jake finished, Jane (140 pounds) tried swinging. How long would it take her to make one complete round trip?

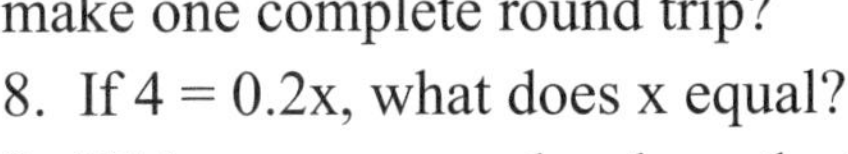

8. If $4 = 0.2x$, what does x equal?

9. Write two numerals where the larger numeral is the smaller number.

10. If $4\frac{1}{2} = (6\frac{1}{4})x$, what does x equal?

The Bridge

from Chapters 1–6

fifth try

1. Stanthony put his regular combination pizza (36 pounds) in the pizza oven. To push it in required 9 pounds of force. What is the coefficient of friction between the pizza and the floor of the oven?

2. While the pizza was cooking, he added 14 pounds of extra cheese. How hard would he have to pull in order to remove the pizza after it was done?

3. The pizza was for Joe. He thought he was very hungry because he had worked hard sitting in his boat and fishing. Joe selected the largest table at Stanthony's. He needed that for a 50-pound pizza. What was the area of the tabletop?

4. Joe liked to play with the melted cheese on his pizza. He put an olive on the end of a 16" string of cheese and swung it back and forth. He then put a heavier olive on the end of that string. Would that increase or decrease the time it took for the cheese pendulum to swing back and forth (assuming that the cheese does not stretch)?

5. If $30\mu = 15$, what does μ equal?

6. Is the weight of a pizza a continuous variable?

7. Is the number of pizzas that Stanthony will cook today a continuous variable?

8. Which is a larger numeral: **5** or 9?

9. Before the pizza arrived, Joe pushed his table closer to the kitchen so that he could watch Stanthony working. Pushing with a force of $5\frac{1}{3}$ pounds he could push the 16-pound table at a constant speed. Find $\mu_{\text{table against the floor}}$

10. If $7y = 5$, what does y equal?

Chapter Seven
Measuring Force

If Fred were Super Fred, he couldn't push that safe down the hallway. However, he could easily pick up the safe and carry it. He carefully examined how strong he was now at the age of 5, and decided that he would have to wait a while in order to develop big muscles.

Fred pressed against the safe with 6.2 pounds of force. The safe didn't move.

Wait a minute! A couple of chapters ago, back on page 30, you said that you would show me how Fred could measure how hard he was pushing or pulling. I don't want to think that you are just making up that 6.2 pounds of force.

Heavens no. I keep the amount of fiction in the Life of Fred books to a minimum.

So how did you know it was 6.2 pounds?

One easy way is to take a scale like this, turn it sideways, and put it between Fred's hand and the safe.

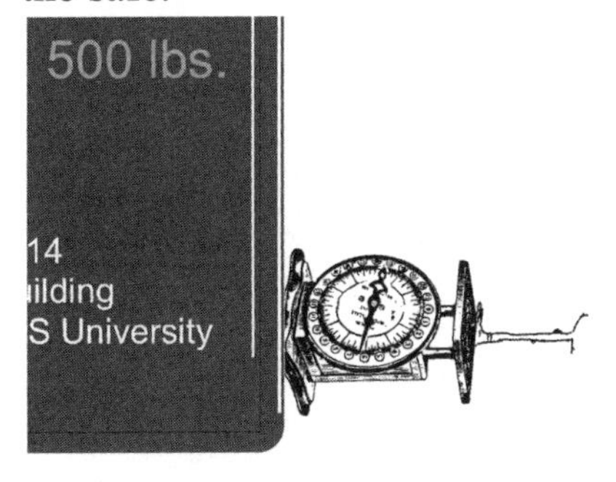

When Fred pushes, I can just read the scale.

But when Fred was pulling his toy safe across his desktop, you can't use that scale! It is only good for measuring a push.

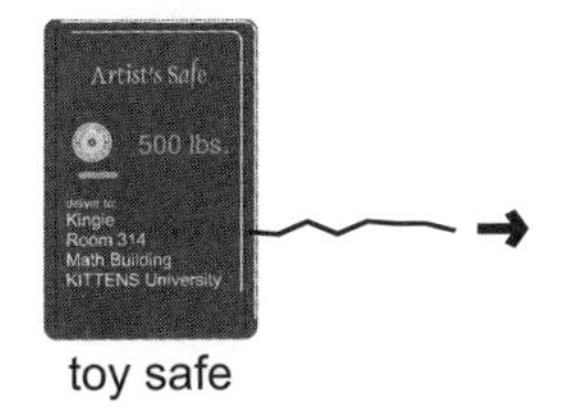

toy safe

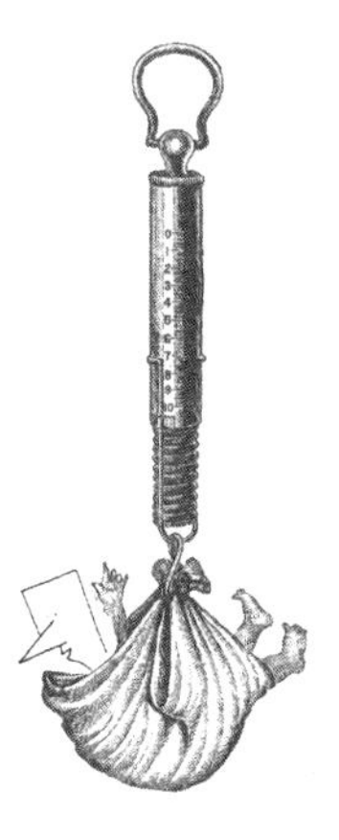

This is a spring scale. It is designed to measure pull and not push. The heavier the object, the more the spring is stretched.

When Fred was born, they wrapped him in a blanket and weighed him on this scale.

Wait! I, as your reader, hate to keep interrupting, but that scale reads seven pounds. Was Fred a normal-sized baby? That is hard to imagine. He is currently five years old and weighs only 37 pounds.

Do you think that those chubby legs and arm belong to Fred? I'll explain that all in *Life of Fred: Geometry Expanded Edition.* For now, I've got to explain how a spring scale works.

Everyone knows that the harder you pull on a spring, the more it stretches. But no one knows how the length of the stretch is related to the amount of the pull, a priori*. Famous saying: *Mathematicians can sit in a comfortable chair with a pencil and a clipboard and invent math. Physicists have to go and shove stuff around.* Or, in this case, they have to pull stuff.

Fred headed back to his office and found a coiled spring toy, the kind that will hop down stairs. He nailed it to the ceiling and attached various weights to the other end.

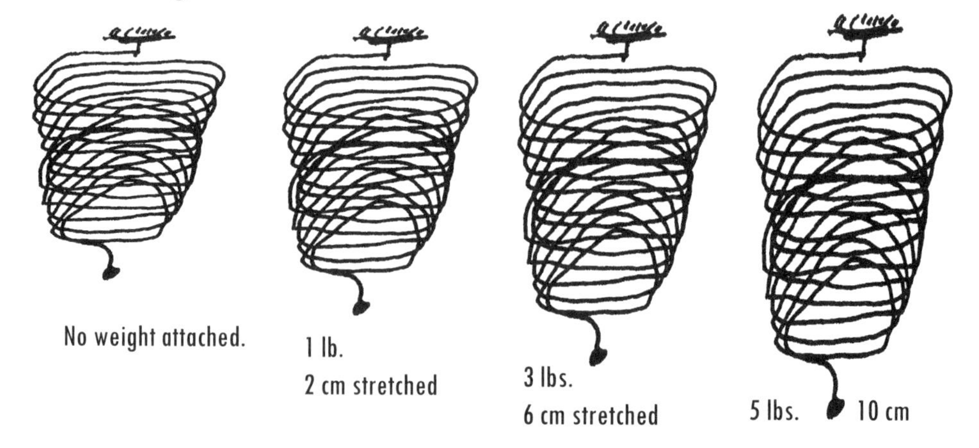

* A-pry-OR-eye. My dictionary lists six other pronunciations of this phrase, such as A-pre-OR-ee. If you know something, a priori, you know it without having to resort to experiment or observation. You already have it figured out in your head.

Fred had done a physics experiment. He didn't know what the results were going to be until he had pushed or pulled stuff around.

Now that he had the results, he could become a mathematician again and figure out what those results meant.

He recorded the results as ordered pairs: (0, 0), (1, 2), (3, 6), and (5, 10). The **first coordinate** represented how much weight was attached to the spring. The **second coordinate** represented how much the spring stretched beyond its normal hanging length.

He plotted those four points.

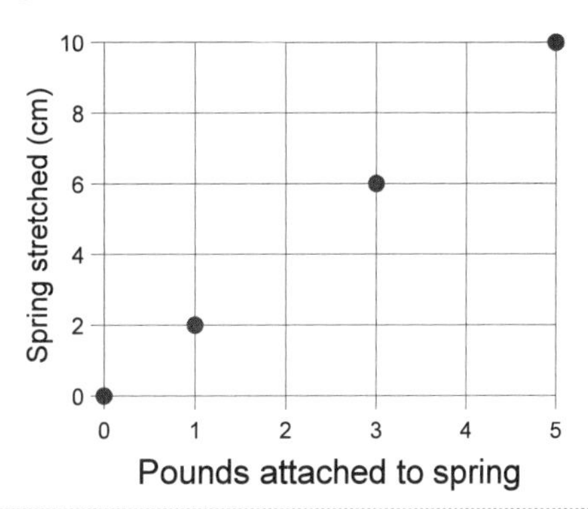

Short review of graphing

To plot (3, 6), you go over 3 and up 6.

(→, ↑)

Your Turn to Play

1. Look at the points on that graph. Which of the following is true?

   A) The points seemed to be scattered at random.

   B) The points seem to lie on a straight line.

2. Guess how much the spring would stretch with a 4-pound weight on it.

3. Fred opened the window and spit. (He made sure that no one was standing below the window before he started this experiment!)

   He recorded ordered pairs: (elapsed time, distance the spit fell).

   At zero seconds, the spit had not fallen at all. (0, 0)

   After one second, it fell 16 feet. (1, 16)

   After two seconds, it fell 64 feet. (2, 64)

Plot these three points.

Are they on a straight line?

.......COMPLETE SOLUTIONS.......

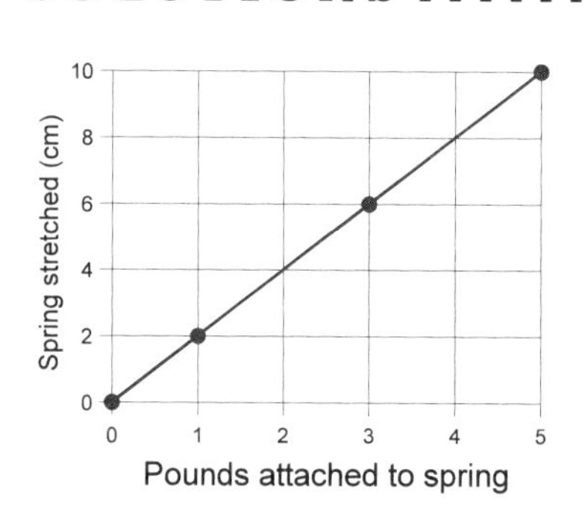

1. B) They seem to be on a straight line.

2. With a 4-pound weight, it looks like the spring would stretch 8 cm.

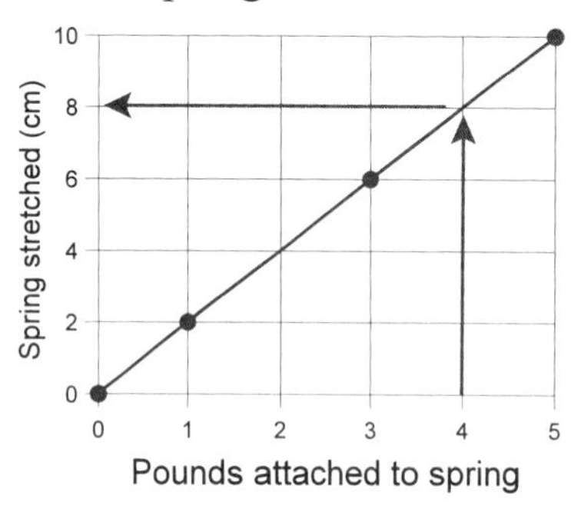

3.

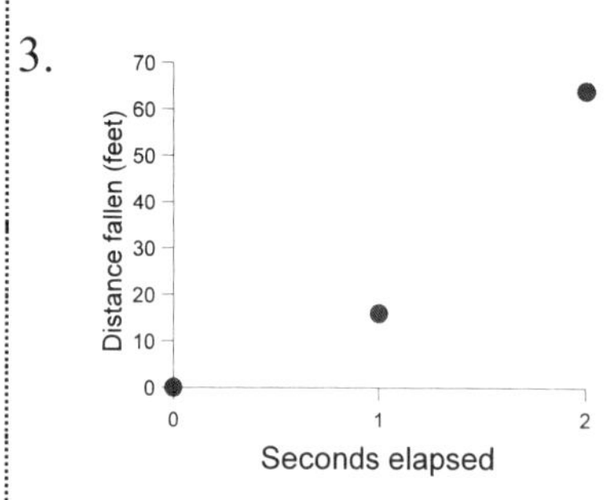

These points are not on a straight line.

If Fred dropped a rock into the Grand Canyon the graph would look like this.

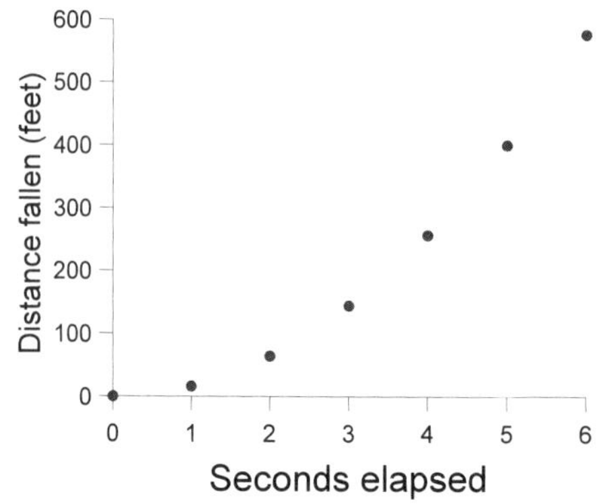

And if you would like the complete truth, if Fred threw a rock into the Grand Canyon, the graph would actually look like this.

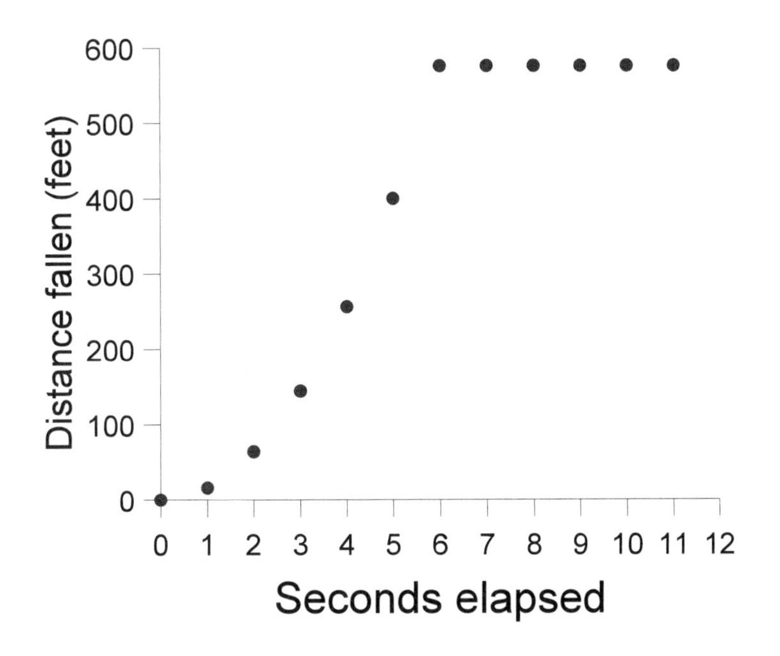

At 6 seconds it had fallen 576 feet. At 7 seconds it had fallen 576 feet. At 8 seconds it had fallen 576 feet.

Okay. I give up. Why did the rock stop falling?

It hit the ground. ☺

Chapter Eight
Absolute Truth

Fred's experiment seemed to point to the conclusion that the force applied to a spring is proportional to how much the spring is stretched. $F = kx$

(F is the force. x is the distance stretched. k is the constant of proportionality.)

Just like the things we mentioned before . . .

✔ If you are driving at a constant speed, the distance you go is proportional to how long you drive. $D = kT$
✔ The wages you receive are proportional to the number of hours worked. $W = kH$
✔ The cost of grapes is proportional to the number of pounds of grapes you buy. $G = kP$
✔ If you swim underwater, the pressure on your eardrums is proportional to how deep you are. $P = kD$
✔ If you are pushing something, the frictional force is proportional to the weight. $F = \mu N$

Stop! I wish that I, your reader, were writing this book. What do you mean by ". . . seemed to point to the conclusion"? Didn't Fred's three experiments PROVE that the force applied to a spring is proportional to how much the spring stretched? The points on the graph were in a straight line.

Fred only attached three different weights to the spring. That doesn't prove anything. It only suggests what might be true.

Okay. How many experiments, how many weights, how many different springs, at how many different temperatures, at how many different places on earth, at how many different times would Fred have to do his stuff before we could stop saying that it just seems to point to the conclusion?

Hee, hee, hee.

Stop your giggling!

I can't help it. You are looking for absolute truth in a physics book. Nothing in a physics book is always true.*

When physicists do experiments, when they push stuff around, they are trying to figure out what might be true.

After they have done a couple of experiments they get a hunch about what might be true.

---

* That is true.

After they do some more experiments and use more expensive equipment, they make a conjecture that something is true.

After their friends (and enemies) also get the same results in their experiments, then their conjecture can turn into a theory.

After those experiments are repeated for centuries, the physicists start to call their theory a law.

Then it is true—when it's a law.

No. Physicists are still just pushing stuff around and guessing what might be true.

It is worse than that. I can't think of a single law of physics that might not some day be proven wrong.

✏ They thought it was a law that atoms were indivisible: they couldn't be broken into anything smaller. ✐✐✐ Then they found protons, neutrons, and electrons inside.

✏ For thousands of years, they thought the sun went around the earth. ✐✐✐ We now think the earth goes around the sun.

✏ For thousands of years, they believed that two different events (for example, one in New York and one in San Francisco) could happen at the same time. ✐✐✐ Einstein convinced people that it is silly to talk about simultaneous events.

✏ Everyone believed the Law of the Conservation of Matter: Matter can neither be created nor destroyed. ✐✐✐ $E = mc^2$.

Your Turn to Play

1. In mathematics, statements that are proved are called **theorems**.
Fill in one word: Physics does not have ____?_____.

Make some guesses:

2. "The earth goes around the sun in an ellipse." Yes or No.
3. "In moving an object at a constant speed, the frictional force is exactly proportional to the weight. $F = \mu N$." Yes or No.
4. "The force applied to a spring is proportional to how much the spring is stretched. $F = kx$." Yes or No.
5. "$2 + 2 = 4$." Yes or No.
6. "There is exactly one star (which we call the sun) in our solar system." Yes or No.

.......COMPLETE SOLUTIONS.......

1. Physics does not have theorems because physics never proves anything. There are two kinds of reasoning: **inductive** and **deductive**.

Inductive reasoning goes from experiments, trials, and observations to tentative conclusions. In its prologue every science book should contain the warning, "These are our best guesses." A lot of what science majors learn in college is outdated ten years after they have graduated.

Deductive reasoning is at the heart of mathematics. Over two thousand years ago, mathematicians proved that in every triangle with two equal sides, there will be two equal angles. It is still true.

implies

2. Almost all the astronomy books state that the orbit of the earth is elliptical. That is close to the truth . . . but it isn't true. The gravitational attraction of the sun isn't the only thing pulling on the earth. All the other planets and the moon also tug on the earth. The earth never makes exactly the same trip around the sun. Each year it is slightly different.

The Three-Body Problem. Suppose you have three things in space and nothing else. Suppose you know the exact position, exact mass of each object in kilograms, and the exact velocity. Can scientists predict where these three things will be a century from now? The surprising answer is that they can't say exactly where they will be. We can only approximate the answers. (For two bodies we can; for three bodies we can't.)

3. and 4. $F = \mu N$ and $F = kx$ are only approximations. They work pretty well for everyday use, but don't move the sliding body too fast and don't pull the spring too far.

5. This is math, not science. It is true.

6. Scientists are not sure! (***Gasp!***)

We know that there is one star in our solar system—our sun—but there may be another star orbiting our sun. This was first proposed in 1984 and given the name Nemesis. We think it might exist because of mass extinctions in the fossil records. (***Double gasp! You mean the fossils and astronomy are connected?!***) Life is stranger than we imagine. I sometimes wonder if it is stranger than we *can* imagine. (That is a famous quote from somewhere.) Look up "Nemesis" on your computer and break the news to your parents. Be gentle. They "know" there is only one sun.

Chapter Nine
Four Ways to Stretch

Fred wasn't the first person to attach weights to a spring and see how far the spring stretched. Three or four hundred years ago (1678) Robert Hooke played with some springs and uttered his famous quote, "*Ut tensio, sic vis.*"

Fred could work in Latin: "*Ut tensio, sic vis.*"

He could work in English: "As the extension, so the force."

Instead, he chose to work in math: F = kx. F is the weight, k is the spring constant of proportionality. x is the distance that the spring is stretched from its starting position. F = kx is called **Hooke's law**.

When he put a three-pound weight on his coiled spring toy, it had stretched 6 cm.

F = kx became 3 = k6.

Since the number is usually written before the letter in algebra, Fred wrote 3 = 6k.

Divide both sides by 6 $\qquad \frac{3}{6} = \frac{6k}{6}$

Simplify each side $\qquad \frac{1}{2} = k$

His experiment allowed him to find the value of k for this spring. Instead of F = kx, he now knew $F = (\frac{1}{2})x$.

He could also have written F = 0.5x if he wanted to work in decimals.

Or he could have worked with the graph he had drawn.

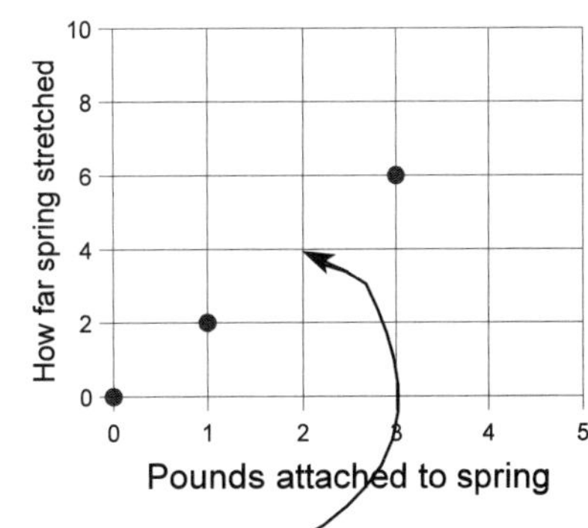

Looking at the graph, he could see that if he attached a two-pound weight to the spring, it would stretch 4 cm.

If he attached a weight that stretched the spring $3\frac{5}{8}$ cm, then

$F = (\frac{1}{2})x$ becomes $F = \frac{1}{2} \times (3\frac{5}{8}) = \frac{1}{2} \times \frac{29}{8} = \frac{29}{16} = 1\frac{13}{16}$ lbs.

Tiny fractions review: To convert $3\frac{5}{8}$ into an improper fraction, you multiply 8 times 3 and add that to the 5.

One of the glories of using the metric system is that you will never see expressions like $3\frac{5}{8}$ cm. Virtually everything is in decimals.

Let's try again. If Fred had attached a weight that stretched the spring 7.3 cm, how heavy was the weight?

We start with	F = 0.5x.
x is 7.3	F = (0.5)(7.3)
Doing the arithmetic	F = 3.65 lbs.

$$\begin{array}{r} 7.3 \\ \times\ 0.5 \\ \hline 365 \end{array} \qquad 3.65$$

It would have been hard to find that exact value by just looking at the graph.

Oops. I shouldn't say *exact* value. F = kx (Hooke's law) is only approximately true for springs.

Stretching Springs—Four Cases

Case #1: Proportional limit.

If you stretch a spring up to its proportional limit, F = kx will be a pretty good approximation of the truth—not exact, but close enough that you can use it for a spring scale.

The proportional limit (2 inches or 2 yards or whatever) will depend on the particular spring.

So how do you tell when you have reached the proportional limit?

That's easy. You draw a graph like Fred did, plotting pounds attached to the spring vs. how far it stretched.

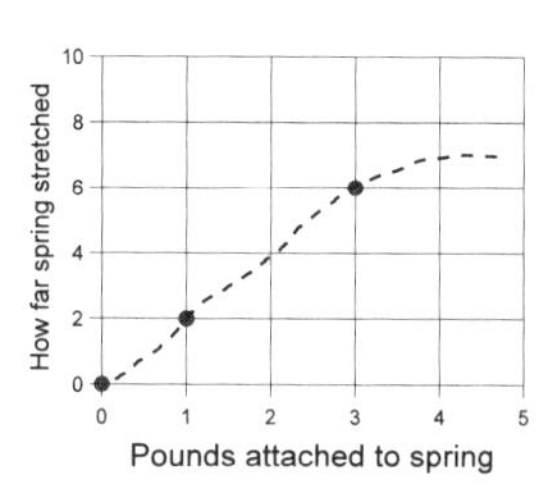

When you get to the spot where the points are no longer in a straight line (collinear), then you have passed the proportional limit.

Case #2: Elastic limit.

Past the proportional limit is the elastic limit. Force (pounds) is no longer proportional to the stretch (inches). But as long as you do not stretch beyond the elastic limit (3 inches or 3 yards or whatever), the spring will snap back to its original shape when you remove the weight.

Case #3: **Plastic region.**

Stretching a spring beyond its elastic limit takes you into the plastic* region. When you take that coiled spring toy that can hop down stairs and stretch it too far, it will be permanently deformed. (Translation: You wrecked it and have to buy another one.)

Case #4: **Breaking point.**

Most metal springs are pretty hard to break. Maybe if we attached an elephant we could reach the breaking point.

One of the easier springs to break is a rubber band.

1. Cars have very large springs in them. Without them (and the tires and seat cushions), driving over a rough road would be agony. Your glasses would bounce off your face. If someone ever tries to sell you solid steel tires, don't buy them.

Four hundred pounds will compress a car spring five inches. Find the value of the constant of proportionality in F = kx.

2. Continuing the previous problem, if you apply 500 pounds to that car spring, how much will it compress the spring?

3. In your experience, do rubber bands have a large or a small plastic region?

4. On this graph identify the proportional limit and the breaking point.

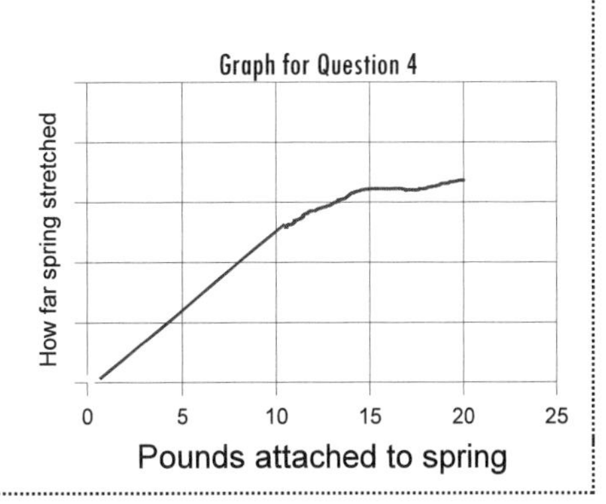

---

* *Plastic* has several meanings: ① the cheap stuff that you buy in the stores, ② credit cards, and ③ able to be molded or deformed. We are using the third definition.

.......COMPLETE SOLUTIONS.......

1. If 400 pounds will compress the spring five inches, then F = 400 and x = 5. F = kx becomes 400 = k5.

Writing the number in front of the letter $400 = 5k$

Divide both sides by 5 $\frac{400}{5} = \frac{5k}{5}$

Simplify $80 = k$

2. If k = 80 and F = 500 pounds,

then F = kx becomes $500 = 80x$

Divide both sides by 80 $\frac{500}{80} = \frac{80x}{80}$

Simplify $\frac{500}{80} = x$

The spring will be compressed $\frac{500}{80} = \frac{50}{8} = \frac{25}{4} = 6\frac{1}{4}$ inches.

3. It's pretty hard (not impossible) to stretch a rubber band so far that it does not return to its original length. Usually when you get just past the elastic limit, the rubber band breaks.

4. In this graph, the amount the spring stretches is proportional to the pounds attached up to about 10 pounds. The line is straight in that region.

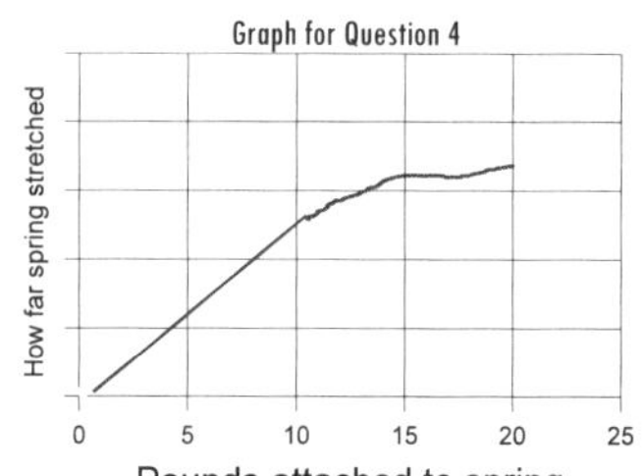

When you get to 20 pounds the line stops. The spring has broken and can "stretch to infinity" at that point.

If I wanted to be silly, I could draw the graph like this:

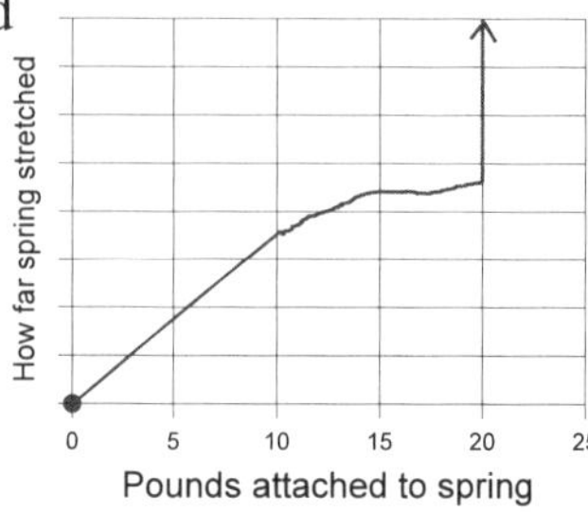

Chapter Ten
Two Kinds of Friction

Fred had pulled his 4-pound toy safe down the hallway at a constant speed. It took 1.7 pounds of force. He had computed mu to be 0.425. He then used $F = \mu N$ and found that $F = (0.425)(500) = 212.5$ pounds of force would be needed to move Kingie's 500-pound safe down the hallway at a constant speed.

Now it was time to put physics to the test. Fred rented a bulldozer that could push with 212.5 pounds of force. Driving it up the two flights of stairs in the Math Building was quite an adventure.

He carefully drove the bulldozer up to the safe so that he would not scratch it. Then he pushed with 212.5 pounds of force.

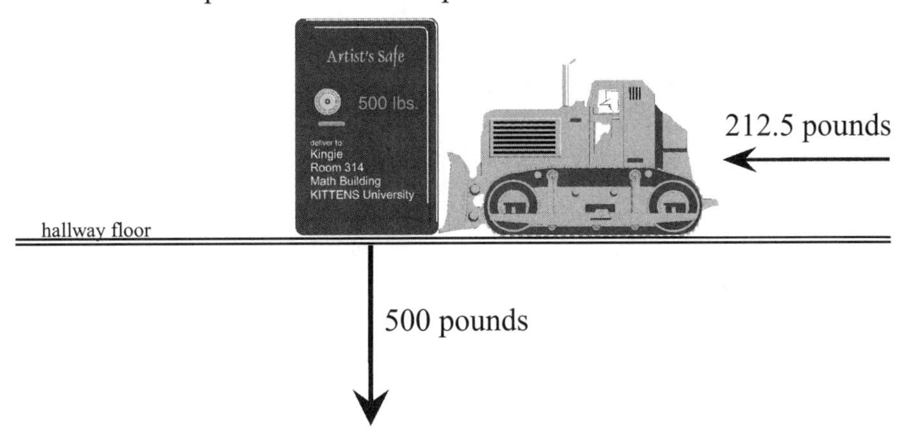

And nothing happened.

The safe did not move an inch.

What had he done wrong? He hopped out of the bulldozer and got his toy safe. He put it down on the hallway floor and pulled it at a constant speed. Had he measured it incorrectly? Scientists can make mistakes in measuring. (So can cooks, dressmakers, surveyors, and doctors.)

1.7 pounds

Fred had not made a mistake in measuring. It still took 1.7 pounds to pull his toy safe down the hallway at a constant speed.

He stopped for a moment and then began pulling again with 1.7 pounds of force. The toy safe didn't move! It just sat there. He gradually increased the force. At 2.4 pounds it moved. Then he could keep moving it down the hallway with 1.7 pounds of force.

Surprise! There are two kinds of friction. Fred didn't expect that. There is a friction for things that are moving, and a different friction for things that are not moving. There are two different kinds of μ.

Friction—Two Kinds

First kind: Kinetic friction.

Kinetic = moving = sliding = in motion.

μ_k is the coefficient of kinetic friction. It is the μ that we worked with in the previous nine chapters. μ_k for the safe and the hallway floor is 0.425. $F = \mu N$ is now $F = \mu_k N$.

One moment! I, your reader, have a question. Why don't they call it μ_m for coefficient of moving friction. I like that a lot better than using k for kinetic.

I do too. But everybody else uses μ_k for kinetic friction. It is a tradition. It is just like the tradition of writing and reading left to right*. If you started writing ƨbɿɒwʞɔɒd it would be harder for others to understand you.

Second kind: Static friction.

Static = stationary = unmoving = at rest.

μ_s is the coefficient of static friction.

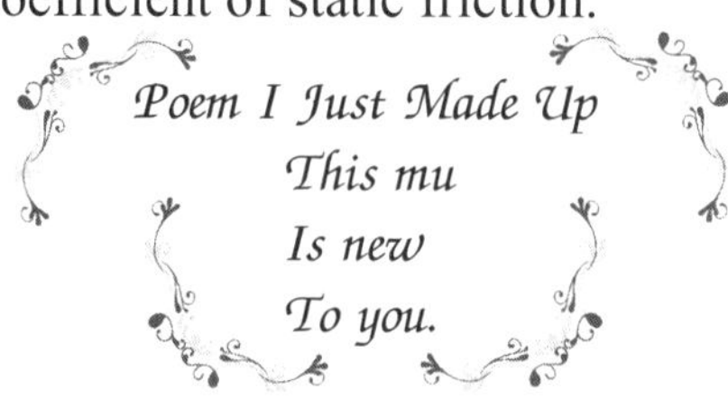

$F = \mu_s N$ is for things that are not moving.

---

* Not all languages are written left to right. Some read right to left and some arrange things vertically.

It is usually harder to get things moving than it is to keep them moving. Usually $\mu_s > \mu_k$.

Often, in everyday life it is harder to start doing something you dread doing than it is to keep on doing it:

❀ Start cleaning up your room than to get the job done once you have begun.

❀ Go to the gym than to work out once you are there.

❀ Sit down and start doing the income taxes than to finish them.

It is getting over the initial hump that is the hard part.

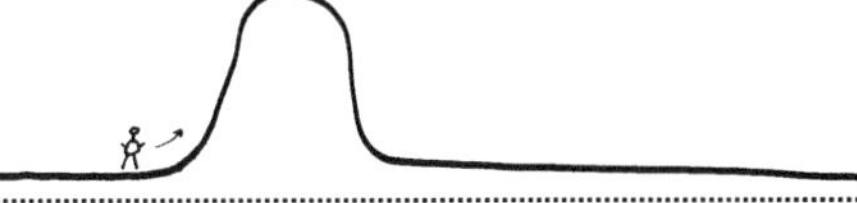

Your Turn to Play

1. Fred's toy safe weighed 4 pounds. It took 2.4 pounds of force to get it moving. Find μ_s.

2. Using the results of the previous problem, find the force needed to get Kingie's 500-pound safe moving on the hallway floor.

3. With Fred's toy safe, it took 2.4 pounds to get it moving. Once it was moving, it took 1.7 pounds to keep it moving at a constant speed.

Make a guess what would happen if Fred kept pushing with 2.4 pounds of force.

4. Rub your hands together and they will get warmer.

Drill into wood and the drill bit will become hot.

Slide a safe down a hallway and the bottom of the safe will get a little warmer.

When you put a piece of bread in a toaster, you send electricity (electrons) through the wires in the toaster. They heat up the wire and toast the bread.

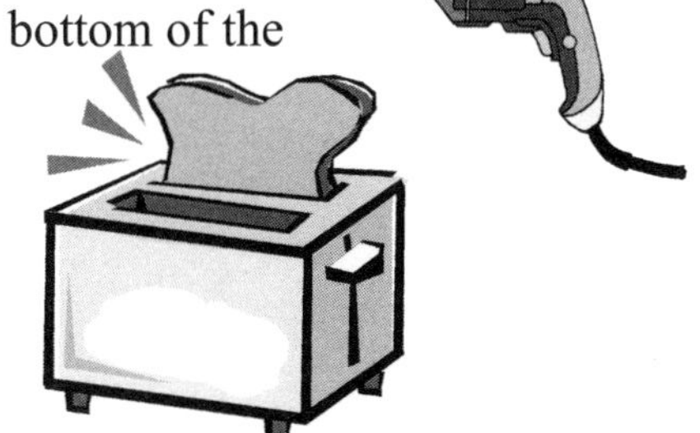

The work of rubbing hands, drilling wood, sliding safes, or running electrons through a wire doesn't get lost.

Friction turns the energy of motion into the energy of ___?___ (fill in one word).

.......COMPLETE SOLUTIONS.......

1. We know $F = \mu_s N$. The toy safe weighs 4 pounds, so N = 4. It took 2.4 pounds of force to get the safe moving, so F = 2.4.

$F = \mu_s N$ becomes $2.4 = \mu_s 4$.

Putting the number in front of the letter	$2.4 = 4\mu_s$
Dividing both sides by 4	$\frac{2.4}{4} = \frac{4\mu_s}{4}$
Simplifying	$0.6 = \mu_s$

2. $F = \mu_s N$ becomes F = (0.6)(500) which is equal to 300.

It will take 300 pounds to get Kingie's safe moving.

3. Since 1.7 pounds of force will keep the toy safe moving at a constant speed, applying 2.4 pounds will make the safe move faster and faster. In the language of physics, the safe will **accelerate**.

In a car the pedal on the right is sometimes called the gas pedal. Another name for that pedal is the accelerator.

4. Friction turns the energy of motion into heat.

Chapter Eleven
Energy

Fred started the bulldozer. He now knew that 300 pounds of force would overcome the static friction and the safe would start moving. He knew that as the safe slid down the hallway floor, friction would turn the energy of motion into heat.

He turned off the bulldozer and thought *Is there a way to turn heat into motion? Or what about other forms of energy?*

He ran back to his office and wrote on his blackboard:

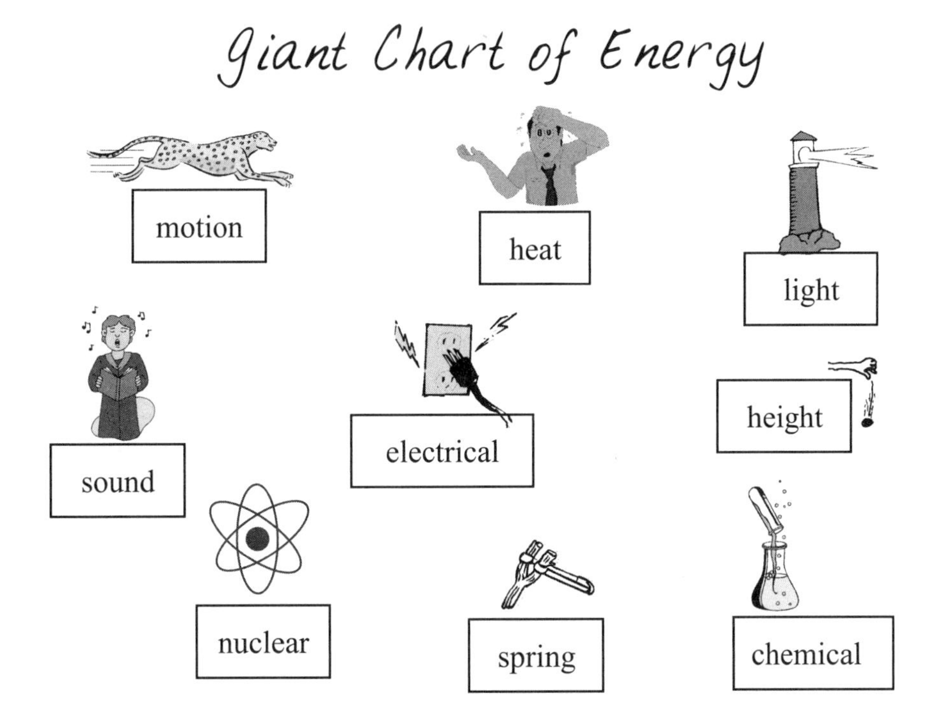

These were all the forms of energy that Fred could think of. He already knew that friction could turn motion into heat. *How many of these forms of energy could be turned into other forms?*

I'll invent a new game and call it the Energy Cards game.

Fred sat down at his desk, took out a piece of paper, and wrote out the rules for the Energy Cards game.

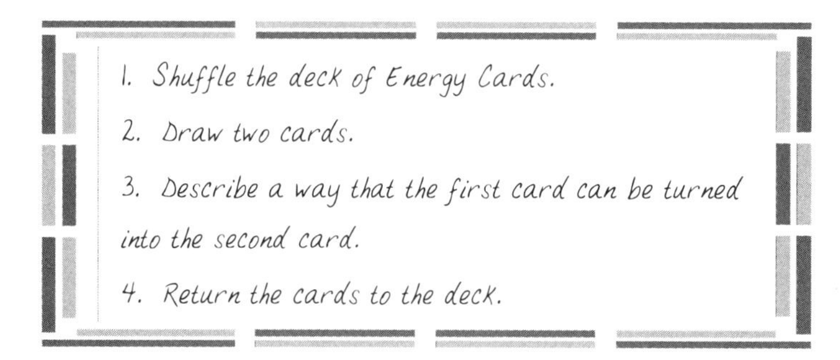
1. Shuffle the deck of Energy Cards.
2. Draw two cards.
3. Describe a way that the first card can be turned into the second card.
4. Return the cards to the deck.

The cards he used were from the chart that Fred had drawn on his blackboard.

Fred shuffled the cards and drew two cards. The first card was

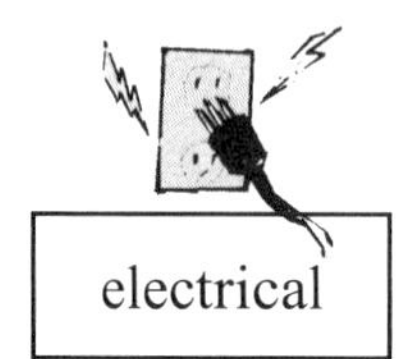

and the second card was

He needed to find a way of turning electrical energy into light energy. That was easy: a lightbulb. Or lightning.

He put the two cards back into the deck and drew two more. This time it was motion and sound. He thought of two possible answers: knocking on wood or playing a harp.

Your Turn to Play the Energy Card Game

1. Turn spring energy into motion energy.
2. Turn height energy into motion energy.
3. Turn electrical energy into sound energy.
4. Turn light energy into electrical energy.
5. Turn electrical energy into motion energy.
6. Turn chemical energy into sound energy. (Hint: Chemical reactions occur in human and animal bodies. Chemical reactions occur when things burn.)

Now some harder ones . . .

7. Turn nuclear energy into heat energy.
8. Turn chemical energy into motion energy.
9. Turn sound energy into electrical energy.
10. Turn motion energy into height energy.
11. Turn chemical energy into light energy.

Fred noticed that four of the energy cards represent ways that energy can be stored for later use.

CHEMICAL: The energy in food can be stored as fat in your body to be used later. Batteries and gasoline store chemical energy.

SPRING: The energy in a jack-in-the-box is waiting to be released.

HEIGHT: A whole bunch of snow at the top of a mountain is energy that can be turned into an avalanche.

NUCLEAR: The energy tucked inside radioactive elements (such as radium and uranium) will release heat over thousands of years.

.......COMPLETE SOLUTIONS.......

1. There are many answers. Yours may be different than mine. Letting go of a stretched rubber band can shoot it across the room.

2. Just let go. It will fall toward the ground.

3. Turn on the radio.

4. Solar panels turn the sun's light into electricity. Solar-powered calculators.

5. This is what electric motors do.

6. Dynamite and firecrackers make a lot of sound when you light them. Gunshots are created from the chemical reactions inside of a bullet.

7. That is what nuclear reactors do.

8. Explosions and bullets are good examples.

9. That is what microphones do.

10. Carry a big rock upstairs.

11. Fireflies and campfires give light because of chemical reactions.

Chapter Twelve
A Time for Action

Fred was having fun with his Energy Cards. There were some combinations that he could not figure out, such as turning sound into nuclear energy.

Kingie looked at Fred playing with his cards. Then he looked out into the hallway. The safe was still sitting out there. He knew that five-year-old boys sometimes get distracted from the job they are supposed to be doing.

It was time for action. Kingie headed out into the hallway. There was his safe. There was a bulldozer. He climbed in, started the motor, and pushed on the accelerator.

The good news was that the safe was narrower than the doorway.

The bad news was that Kingie had never driven a bulldozer. He tore out half of the doorway pushing the safe into the office.

The good news was that the safe was now in the office.

The bad news was that the bulldozer was now in the office.

The good news was that Kingie knew that he wasn't supposed to leave a bulldozer in Fred's office. He drove it back out into the hallway.

The bad news was that he accidentally tore out the other half of the doorway.

Instead of 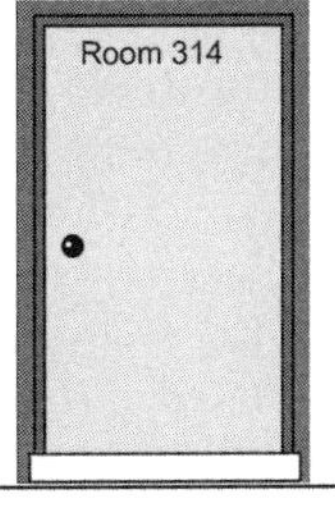

Fred's doorway now was

Happily, Kingie decided not to drive the bulldozer down the hallway. He would probably have destroyed the nine vending machines.

Kingie hopped out and headed to the janitor's office. His office was easy to find. There was a big flashing sign in front of it.

SAMUEL P. WISTROM

Educational Facility Math Department Building
Chief Inspector/Planner/Remediator
for offices 225–324

It was a fancy sign that indicated that he was the janitor for half of the building.

Kingie knocked on the door. He could hear Samuel's loud television set. He knocked again. Samuel opened the door.

"Hi, Sam," Kingie said. "We have a little problem with the door to our office. Can you do a little fix up?"

"I'd be glad to help. What seems to be the problem? Is the door sticking or is it squeaking ?"

"No," Kingie answered. "Neither of those. You'll have to see it to understand what needs to be done."

"Okay. Is there any hurry? I'd like to finish watching my show first. It's 'Peter Puppet Plays in the Peanut Palace.' It's really funny."

Kingie said, "That would be fine. There is no hurry," and headed back to see Fred.

When Kingie got back to the office, Fred was on his computer trying to find companies that would manufacture his Energy Cards game. He was going to get bids from three different companies to produce 2,000 copies of his game. He had high hopes that lots of people would love to play his game.

Fred said to Kingie, "I'll get your safe moved in just a moment."

"There's no need," Kingie told him. "I did it while you were working on your new game. I just used the bulldozer."

"I didn't know that dolls could drive bulldozers. Who taught you?"

Kingie smiled. "No one. I just learned as I did it. By the way, I also made it easier for students to come and see you."

Fred didn't know what Kingie meant until he stopped concentrating on his computer and looked at the doorway. Fred became very silent. He had a beautiful view of the hallway that he didn't have before.

He knew that by law he was responsible for any damage that his doll did. He fainted.

Your Turn to Play

1. Fred's salary for teaching at KITTENS University was $500 per month. How long will it take Fred to earn enough to pay for the $1,200 in damages?

2. Fred fell on the broken door that lay on the ground. That didn't look very comfortable, so Kingie pushed the door (with Fred on top of it) toward Fred's sleeping bag.

Fred weighed 37 pounds. The door, 13 pounds. Kingie had to push with a force of 21 pounds in order to get the door moving with Fred on it. What is the coefficient of static friction (μ_s) between the door (with Fred on it) and the office floor?

3. Once Kingie got it moving, it only took 15 pounds to keep the door and Fred moving at a constant speed. What is the coefficient of kinetic friction between the door and the office floor?

4. When Kingie got Fred near the sleeping bag, he lifted up the edge of the door. Fred slipped off the door and into his sleeping bag. Kingie then slid the door out of the office and into the hallway.

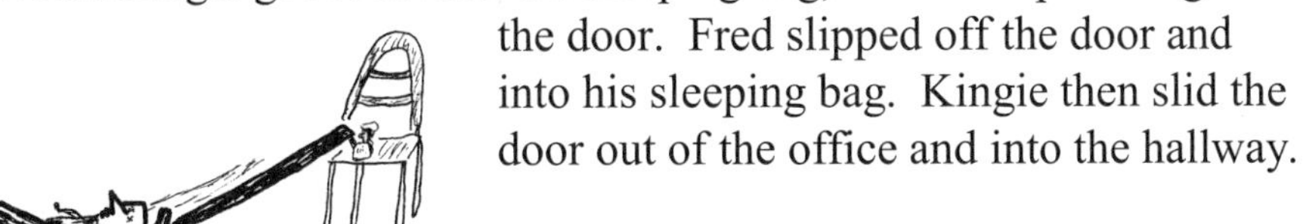

How much force would it take Kingie to get the door (without Fred on it) moving?

How much force would it take to keep the door moving at a constant speed?

.......COMPLETE SOLUTIONS.......

1. How long would it take to pay $1,200 at the rate of $500 per month? If you don't know whether to add, subtract, multiply, or divide, restate the problem using very simple numbers. Suppose he needed to pay $10 and he earned $2 per month. It would take 5 months. You divided.

So the original problem becomes

$$\begin{array}{r} 2.4 \\ 500\overline{)1200.0} \\ \underline{1000} \\ 2000 \\ \underline{2000} \end{array}$$

It will take Fred 2.4 months to pay for the damage.

2. The weight was 50 pounds (37 + 13). The force was 21 pounds.
$F = \mu_s N$ becomes $21 = \mu_s 50$.

Put the number in front of the letters $21 = 50\mu_s$

Divide both sides by 50 $\frac{21}{50} = \frac{50\mu_s}{50}$

Simplify both sides $0.42 = \mu_s$

$$\begin{array}{r} 0.42 \\ 50\overline{)21.00} \\ \underline{200} \\ 100 \\ \underline{100} \end{array}$$

3. $F = \mu_k N$ becomes $15 = 50\mu_k$

Divide both sides by 50 $\frac{15}{50} = \mu_k$

Simplify $\frac{3}{10} = \mu_k$

Did you see that! I skipped a step.

To reduce $\frac{15}{50}$ divide top and bottom by 5.

Or you could work in decimals and get 0.3.

4. From problem 2, we know that μ_s is 0.42.
$F = \mu_s N$ becomes $F = (0.42)(13) = 5.46$ pounds to get the door moving.
From problem 3, we know that μ_k is 0.3
$F = \mu_k N$ becomes $F = (0.3)(13) = 3.9$ pounds to keep the door moving.

The Bridge

from Chapters 1–12

first try

Goal: Get 9 or more right and you cross the bridge.

1. Joe liked to go fishing in the Great Lake at the edge of the KITTENS campus. Every fish that he has caught in the Great Lake has been gray. He now claims that every fish in that lake is gray. Is this an example of inductive or deductive reasoning?

2. Joe thought that Hooke's law was, "It is easier to catch a fish if you use a hook." Joe was confusing Hooke's law with Hook's law.

State the equation for Hooke's law.

3. If $77 = 13\mu$, what does μ equal?

4. Joe had just purchased 345 meters of new fishing line. He wound it onto his fishing reel at the rate of 15 meters per minute. How long would it take him to finish that job?

5. His new fishing line was made from a combination of rubber bands, spaghetti, and calamari.* His new line was a lot more stretchy than his old wire line.

It wasn't long before Joe hooked a "fairly large" fish. As Joe reeled in the fish, which of these four should he have been worried about with his new line?

A) the proportional limit of the line
B) the elastic limit
C) the plastic region
D) the breaking point

This is how Joe imagined it.

6. When Joe told Darlene about the giant fish that he had caught, he called it the great gray whale. He named it Moby Richard. When she asked him where this fish was, Joe said that he had weighed it and thrown it back into the lake. The spring that Joe had used would stretch $3\frac{1}{3}$ inches with a 5-

---

* Years ago, my girlfriend and I were at an Italian restaurant. She urged me to try the spaghetti with calamari. I ordered it. I didn't know what calamari was. If I had been Fred Gauss, I probably would have fainted when it was served.

The Bridge
from Chapters 1–12

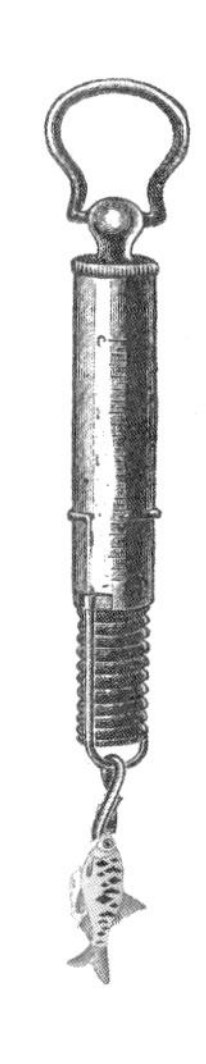

ounce weight on it. What was the constant of proportionality of the spring? (There is no need to convert into pounds. Just work in ounces.)

7. Joe told Darlene that with Moby Richard attached the spring stretched 2 inches. How much did the fish weigh?

8. When Joe threw the fish back into the lake, one of the seagulls dove and ate it. Joe started counting the number of birds circling overhead. Was that number a continuous variable?

9. Years ago, Darlene gave Joe a comic book entitled *Moby Richard*. Joe read the first sentence: "Call me Fish Meal." He thought that was very funny. That is as far as he read. The cover was 8" x 10.5". What was the area of the cover?

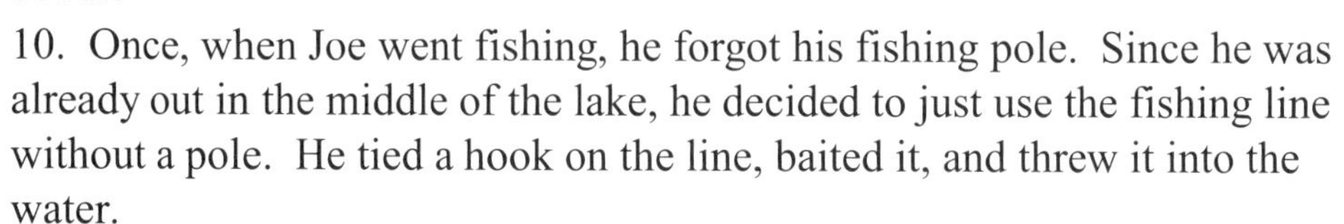

10. Once, when Joe went fishing, he forgot his fishing pole. Since he was already out in the middle of the lake, he decided to just use the fishing line without a pole. He tied a hook on the line, baited it, and threw it into the water.

Joe missed the water and hit a passing speed boat. (Error #1) He held tightly (20 pounds of force) as the line slipped through his hands. (Error #2) The speedboat was pulling the line through Joe's hands at a constant speed. His hands started to feel very warm. Joe didn't relax his grip. (Error #3)

Which of the nine forms of energy we have mentioned was being converted into which other form?

Alphabetical List of some Forms of Energy

1. chemical
2. electrical
3. heat
4. height
5. light
6. motion
7. nuclear
8. sound
9. spring

(Many readers don't get 90% right on the first try. That's why there are four more tries.)

The Bridge

from Chapters 1–12

second try

1. Darlene subscribes to *Immense Bridal Dresses* magazine. It measures 14" x 23" so that the reader can view the details of every color photograph. What is the area of the cover?

2. One picture showed a bridal dress with the train* of the dress so long that a flower girl could stand on it and be pulled along by the bride.** The bride would have to pull with a force of 28 pounds to drag a 42-pound flower girl. What is the coefficient of friction between the dress and the ground?

3. If $24 = 30x$, what does x equal?

4. The current issue of *Immense Bridal Dresses* weighs 12 pounds. It took 4 pounds of force to get it moving when Darlene tried to pull it toward her on the kitchen table. Find the coefficient of static friction between the magazine and the kitchen table.

5. The magazine fell off the table and made a loud crash as it hit the floor. Which *three* of the nine forms of energy we have mentioned were involved?

6. Is the speed at which the magazine fell a continuous or a discrete variable?

7. She was planning to make her bridal veil out of rubber. Pulling on the bottom of the veil with a force of 4.8 pounds would stretch it 3 inches from its normal length. How far would it stretch if she pulled it with 0.8 pounds of force?

8. Graph (100 guests, \$3,000), (200 guests, \$5,000).

9. If you know that $\mu_s = 0.6$, can you determine from that the value of μ_k?

10. If $A = \ell w$, what does ℓ equal?

---

* The train of a dress is the back end that trails behind on the ground.

** Normally, the flower girl goes ahead of the bride.

The Bridge

from Chapters 1–12

third try

1. Fred was reading Christina Rossetti. He was delighted when he came to the line: ***Our teachers teach that one and one make two. . . .**** *

A poet for mathematicians Fred thought. He imagined going on a picnic with her. He could slide the 15-pound picnic basket that he made over to her side of the blanket that he had spread on the ground. It would take 5 pounds of force to keep it moving at a constant speed. Find the coefficient of kinetic friction of the basket on the blanket.

2. 

A cute little 2-pound alligator hopped into the basket. How much force would it now take to push the basket?

3. Fred imagined that he would hand Christina a card with some math on it. (What else is there to do on a picnic?) The card would measure 5.2 cm by 7 cm. What is the area of that card?

4. Is the number of alligators in a picnic basket a continuous variable?

5. Christina took the alligator and set it on the ground. It headed straight toward the Great Lake. In 2 minutes it was 76 feet away. How fast was it running?

6. She turned the card over and wrote: *1 8*. Which numeral was larger?

7. She continued writing *1 8 3 0*. She explained to him that that was the year she was born. Fred was impressed that her handwriting was so good for someone that old. She told him that she kept fit by exercising. She took a spring out of her purse and with 16 pounds of force stretched it 10 inches from its normal length. (continued next page)

---

* From Rossetti's poem "Later Life #16." Fred would not have understood the next line of that poem: ***Later, Love rules that one and one make one. . . .*** When you are five years old, there are things you don't yet understand.

The Bridge

from Chapters 1–12

Fred tried. With 10 pounds of force, how far could Fred stretch that spring?

8. Christina took a red cotton napkin out of her purse and placed it on the blanket. She took a blue cotton napkin and handed it to Fred. She took a yellow cotton napkin and tossed it to the alligator. She took a purple cotton napkin and polished one of the apples. She had many more napkins in her purse.

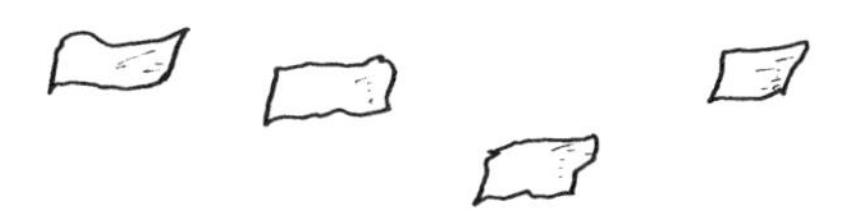

Fred made the guess that all of the napkins that Christina had were made of cotton. Is this an example of inductive or deductive reasoning?

9. If $80y = 3$, what does y equal?

10. Is the speed of light exactly 299,792,458 meters per second or only approximately 299,792,458 meters per second?

The Bridge
from Chapters 1–12

fourth try

1. Jake, Jack, and Jane were three sailors visiting New York City. One spot that they didn't want to miss was Central Park. It is a rectangle 0.5 miles by 2.5 miles. A big park! It is similar in size to San Francisco's Golden Gate Park. What is its area?

2. There are 21 playgrounds in the park. The three sailors headed to the largest one (Heckscher Playground). Jane (140 pounds) picked out a large swing and swung back and forth without pumping. Every 3 seconds she completed one round trip. Then Jack (155 pounds) took a turn on the same swing. Was the time of his round trip greater than, equal to, or less than 3 seconds?

3. They were playing tag in one of the large meadows. Jake was running and slipped on the grass. He slid for several feet before coming to a stop. His white pants now had a large green stain on the back. As he was sliding, which coefficient of friction—μ_s or μ_k—was the one that applied?

4. If 52 = 13x, what does x equal?

5. Jake (155 pounds) lay on the lawn. Jane grabbed one of his legs and pulled him. It took 31 pounds of force to start Jake sliding again. He now had an even larger green stain on his pants.

Jack (180 pounds) was sitting on the grass playing with his cell phone. Jane grabbed one of his legs and pulled him also. How hard would she have to pull to start him sliding? (Assume Jake and Jack have the same coefficients of friction.)

6. When they stood up, they noticed that Gauss Movie Studios had been filming them. Jake hoped that they hadn't gotten a picture of the green stains on his pants.

Facts: Central Park is the most filmed spot in the entire world. In 1908 Vitagraph Studios shot the first film of Romeo and Juliet ever produced in America in Central Park. Through 2011, over three hundred films have been made using scenes in the park (including When Harry Met Sally, Kramer vs Kramer, and Home Alone 2).

Is the number of films that uses scenes in Central Park a discrete variable?

The Bridge

from Chapters 1–12

7. Jane asked the director of the film if he was hiring actors. She had always wanted to star in a movie. Would Jane's pay be a continuous or discrete variable?

8. The director told Jane that she could report tomorrow to be a part of a mob scene. Jane was unhappy because they had to report back to the ship by 10 p.m. that night.

Meanwhile, Jake was rubbing the green stains on his pants with his handkerchief. The energy of motion (kinetic energy) was being transformed into what other form of energy?

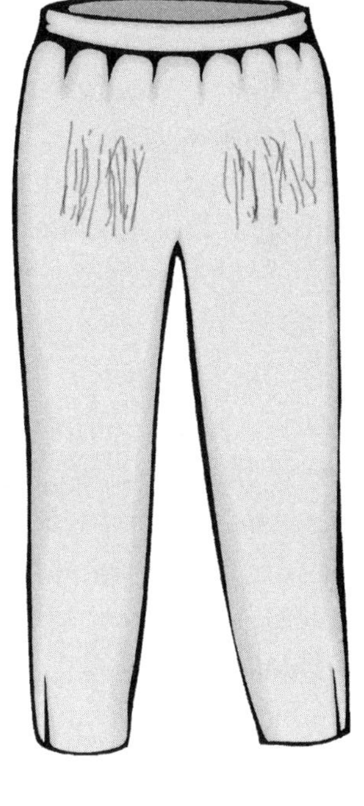

9. Jane knew that rubbing grass stains with your handkerchief would not work. She knew because she had tried it hundreds of times when she was a kid. Is this an example of deductive or inductive reasoning?

10. If $\left(\frac{2}{3}\right)x = 6$, what does x equal?

The Bridge
from Chapters 1–12

fifth try

1. PieOne was noted for its unusual pizzas. Stanthony would sometimes add rice as a topping. Every time he did that, the customers complained. He concluded that he should never use rice as a topping. Is this an example of deductive or inductive reasoning?

2. Stanthony bought a springy pizza flipper. When he tried to flip a 52-pound pizza, the flipper got bent out of shape. He had to throw it away. Which of these four applies to that flipper? 1) It had reached its proportional limit; 2) It had reached its elastic limit; 3) It had entered the plastic region; 4) It had passed the breaking point.

3. If 18y = 24, what does y equal?

4. Graph (20 pizzas, $50 profit), (40 pizzas, $100 profit).

5. When larger parties arrived at PieOne, Stanthony would sometimes have to push several tables together. It took 40 pounds of force to get a 60-pound table moving. Find μ_s.

6. From the information in the previous problem, can you determine the value of μ_k?

7. One of the special pizzas was called the Table Top Pizza. It was a rectangle that measured 50 inches by 70 inches. What is the area of that pizza?

8. What force would be needed to push an 8-pound chair across the room at a constant speed if μ_k was equal to 0.7?

9. Is the number of customers at PieOne a continuous variable?

10. Which of these numbers is larger: 3 or 8?

Chapter Thirteen
Flabbergasted Fred

In a minute Fred awoke. He looked at the doorway and was speechless. Kingie thought that this was the perfect time to try to distract Fred from the doorway disaster.

"I can do something you can't do," Kingie began.

Fred was silent. He knew Kingie had more to say.

"Do you remember how you found the coefficient of static friction (μ_s) with your toy safe against the hallway floor?"

"Sure," Fred answered. "I used a spring scale and found that it took 2.4 pounds to get my 4-pound safe moving. I put $F = \mu_s N$ in the form $\frac{F}{N} = \mu_s$ and then put in the values for F and N. I did the division and out popped the value for μ_s."

"What if I told you that I could find μ_s without using any scale to find the amount of force necessary to get the safe moving?" Kingie asked.

"I would say that's silly. You need both F and N in order to find the value of μ_s."

Kingie continued, "What if I told you that I could find μ_s for your safe against the hallway flooring without a scale and without knowing the weight of your safe?"

"Impossible!" Fred exclaimed. Fred looked around and found a piece of the hallway flooring that Kingie had gouged out with the bulldozer. He handed it to Kingie. Then he reached into a desk drawer and got out another toy safe. This one was a little smaller than the 4-pound safe.

He handed the safe to Kingie and said, "Okay. No scale and you don't know the weight of this safe. Find μ_s and flabbergast me."

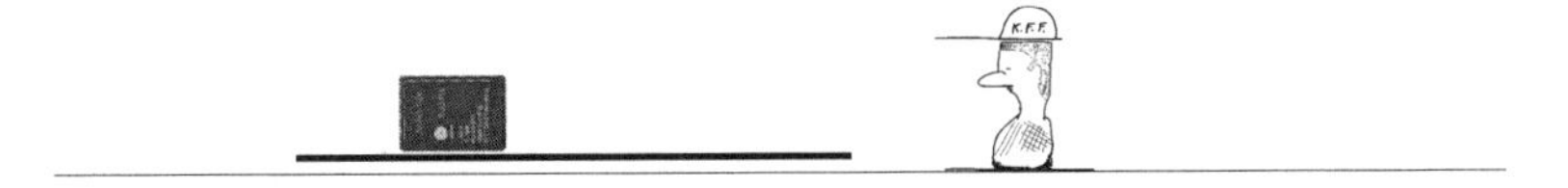

Fred already knew that μ_s was 0.6. (This was found in problem 1 at the end of Chapter 10.) He had used a scale and had known the weight of the safe. He could not imagine how Kingie was going to do this magic.

"If you would, please lift up the left end of that flooring until the safe just starts to slide," Kingie said. He got this idea when he had used the door to slide Fred into his sleeping bag.

Fred, who is bigger and stronger than his doll, complied.

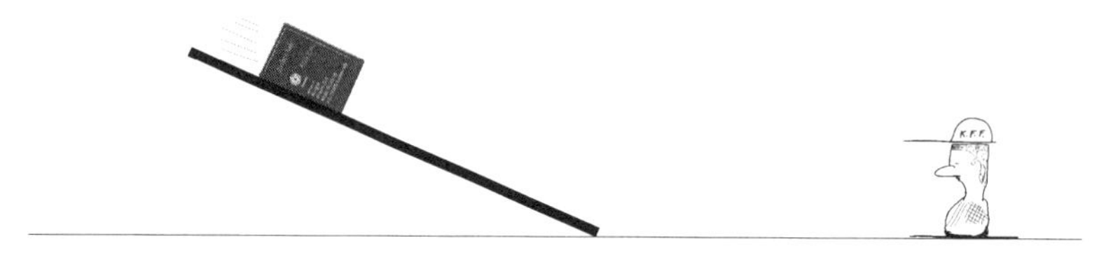

"We are almost done," Kingie announced.

Fred thought *Kingie is just fooling around. We haven't "done" virtually anything yet.*

"Please make two measurements," Kingie told Fred, pointing to the places he wanted measured.

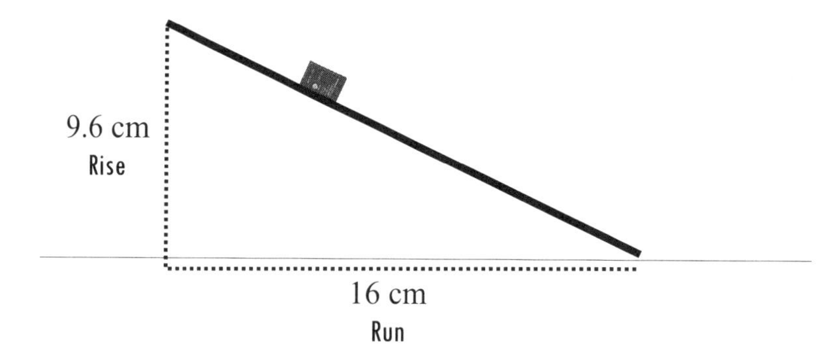

"Now divide the rise by the run," Kingie instructed Fred.

"That will give you the slope of the line!" Fred almost shouted.

"Just do it," Kingie told him.

Fred did the division on a piece of scratch paper. When he read the quotient—0.6—he trembled.

"μ_s is 0.6," Kingie declared with a big smile.

```
     0.6
16) 9.6
    9 6
```

Fred's work

Fred said, "No, no, no, no, no, no, no. This must be some trick."

Kingie said, "Yes, yes, yes, yes, yes, yes, and it is no trick. There are some very, very, very surprising things in physics, just like there are in math."

"But you don't know the weight of the toy safe. How can you find the coefficient of friction?" Fred was floored.*

Your Turn to Experiment

Finding μ_s Using the Kingie Method**

This is, perhaps, your very first physics experiment. A small word of warning needs to be made as you enter the world of real physics: You are going to be *doing* things rather than just reading.

Everyone knows that being a mathematician is one of the safest occupations. You sit at a desk with a clipboard, a pencil, and a strawberry milkshake. The worst thing that can happen is that you knock over the milkshake, but that is rarely fatal.

If you are a lumberjack, you are more likely to be injured.

If you sell hot dogs, you might get poked.

And do we have to mention what could be dangerous about being a tiger tamer?

---

* To be floored is to be surprised, confounded, shocked, overwhelmed, have your socks knocked off, and be blown out of the water.

** For the time being, we are going to trust Kingie. He has said that the slope of the line, which is equal to $\frac{\text{Rise}}{\text{Run}}$ when the safe just starts to slide is equal to the coefficient of static friction, μ_s. In the next chapters we will show that he is correct.

So as you do this physics experiment, consider carefully whether you will need goggles, heavy gloves, and an apron. It is up to you to be safe.

The Experiment

First, find a flat surface that can be tilted. It will not be necessary to tear up the floor as Kingie did.

Second, find something that will slide on that surface. For example, a coin, a rubber band, a paperclip, or a glove. Marbles are no good. They roll. Snot is no good. It won't slide.

Third, tilt the surface until the thing starts to slide. Measure the Rise and the Run.

Last, divide the Rise by the Run to obtain μ_s.

That looks like fun. Here is what I found around my house:

Flash drive on a Fred book

Clothespin on cardboard box cover

This didn't work. I tried lifting a building to tilt the floor.

Physics can be dangerous. Big knife on a binder

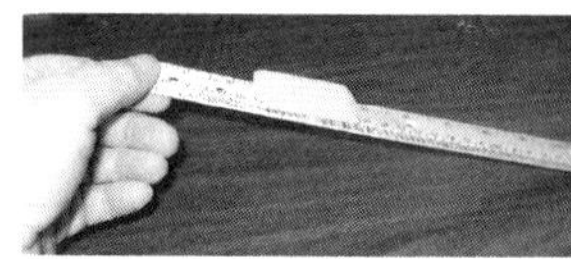
Eraser on a ruler

No other physics book has done a can on a griddle.

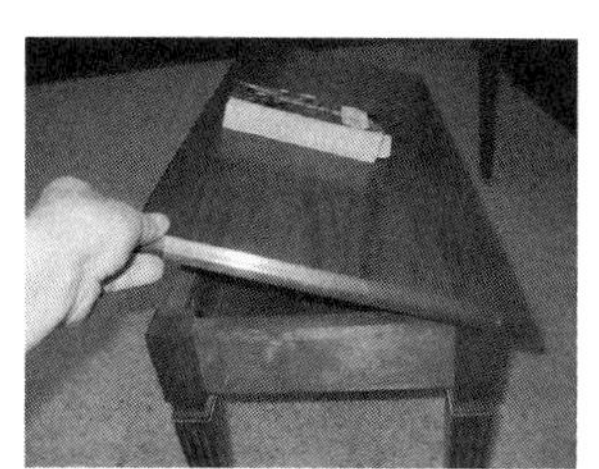
Book on a bench

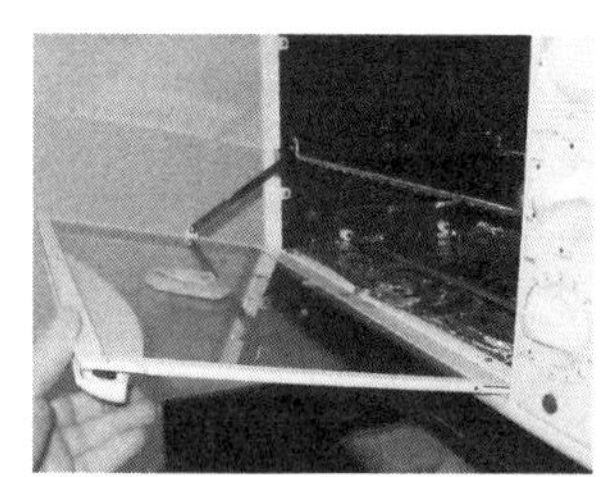
Cookie on a convection oven door

Calculator on a clipboard

Chapter Fourteen
Normal Force

It was time for Kingie to explain. Fred knew that there were certain things that you had to trust other people about.* But this was not one of them. Fred had no idea why the slope ($= \frac{\text{Rise}}{\text{Run}}$) should exactly equal the coefficient of static friction ($\mu_s = \frac{F}{N}$). Those things seemed so unrelated.

Kingie Explains

Once upon a time everything that we slid was on a level surface.

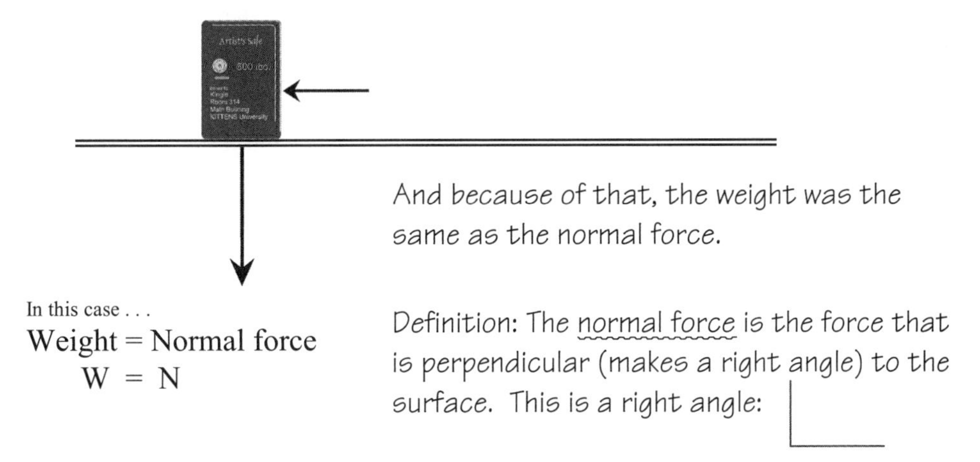

And because of that, the weight was the same as the normal force.

In this case . . .
Weight = Normal force
W = N

Definition: The normal force is the force that is perpendicular (makes a right angle) to the surface. This is a right angle:

---

* With your own eyes, have you ever seen the North Pole? Someone told you that it exists. Photographs don't count. I could show you a picture of the West Pole.

Official picture of the West Pole

How do you know that China exists? That you have a liver? That Martin Luther ever lived? That the stuff you read about in the newspaper is true? 99.99315% of the things we "know" are things we take on faith because others have seen them and have told us about them.

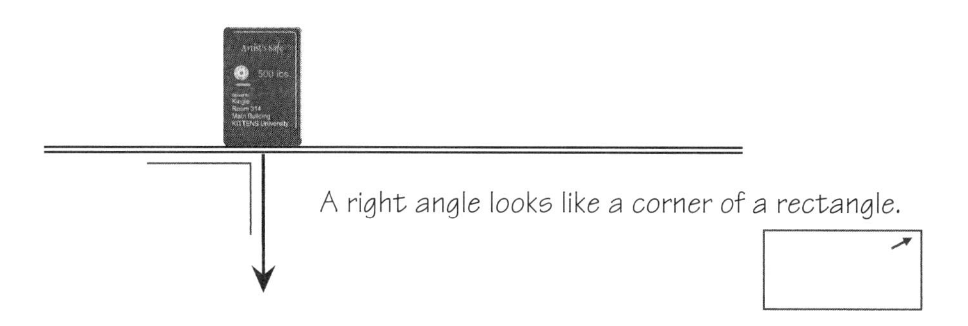

When the surface is level, the full weight of the safe is pressing against the surface.

If we tilt the surface, the safe doesn't press as hard against the surface.

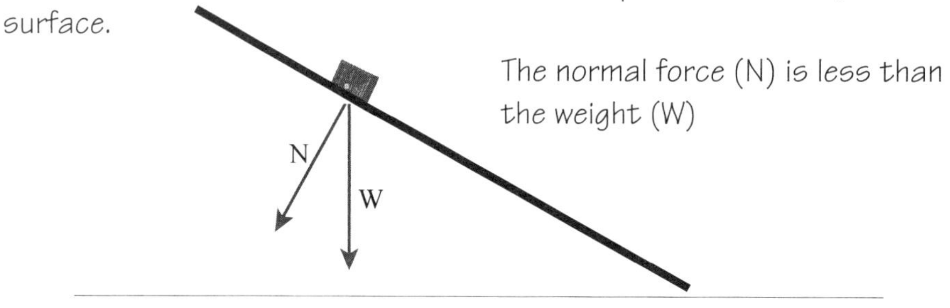

If you really tilt the surface, the weight of the safe stays the same, but the normal force—the force of the safe against the surface—is much less.

With very little normal force, there is very little friction, and the safe will slide.

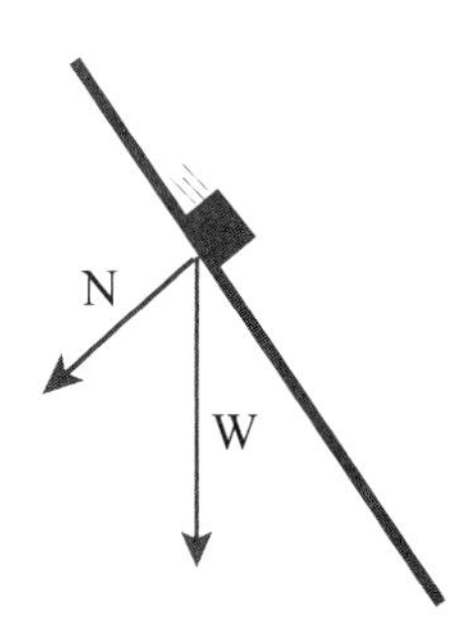

Are there any questions so far?

Fred raised his hand. "I can see how you draw the tilt of the surface. I can see how you draw the arrow for the weight. (It is straight down and the length of the arrow represents the 500 pounds that the safe weighs.) I can see how you draw the direction of the normal force. (It is at right angle to the surface.) But what I don't understand is how long to make the arrow for the normal force."

Good question. I hope that you are good at art because finding the length of the normal arrow is done by drawing perpendicular lines. That shouldn't be too hard since your head has lots of perpendicular lines.

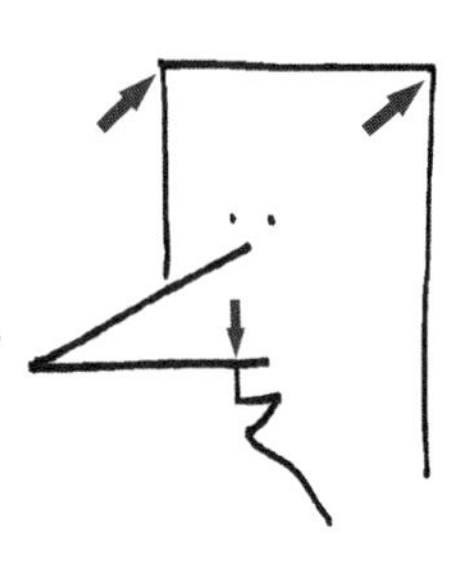

We start with two things: (1) the tilt of the surface and (2) the weight of the safe.

The weight of the safe is represented by a downward arrow. In physics that arrow is called a vector. Vectors have length and direction. The length of this weight vector tells you how much the safe weighs. The direction of the vector tells you which way gravity is pulling the safe.

We know what the direction of the normal vector will be. It is at a right angle to the surface. We just don't know how long to draw it.

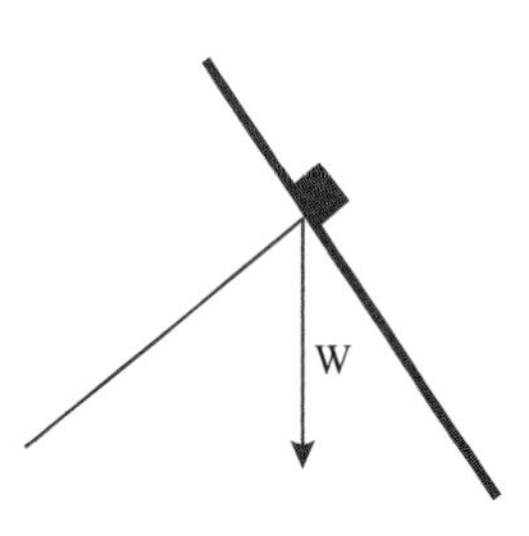

Here comes the magic. Are you ready?

I'm ready.

Make a box.

You mean a rectangle?

Yes. Make a rectangle. That's all you have to do. Now you have the normal vector.

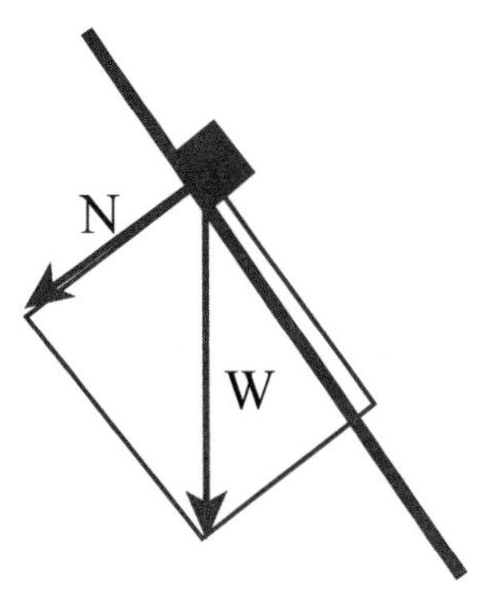

The weight (W) creates a force (N) against the surface. The length of N is less than the length of W (unless the surface is level).

The weight of the safe actually creates two forces: the normal force against the surface and the force that makes the safe want to slide down the surface. Friction is keeping the safe from sliding.

In the language of physics, W is resolved into N and F.

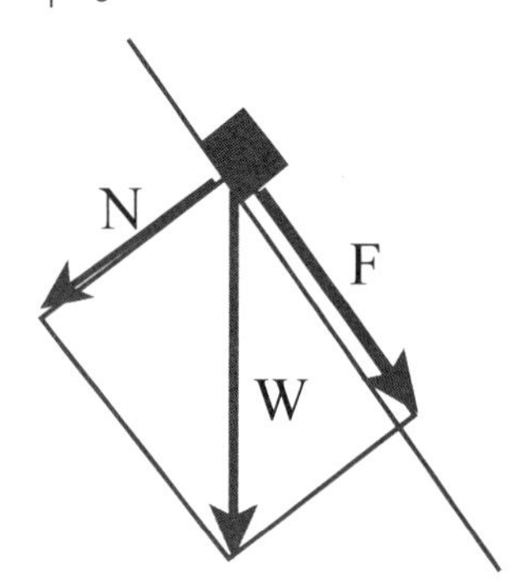

Fred raised his hand again. "You mean that you are adding arrows together?"

In physics they are called vectors.

Okay. Do you mean that in physics, you are adding vectors together? I thought you could only add numbers together.

Watch me. N + F = W. I did it.

I can do more tricks. Recall that a vector has two parts: its direction and its length. So all these vectors are equal.

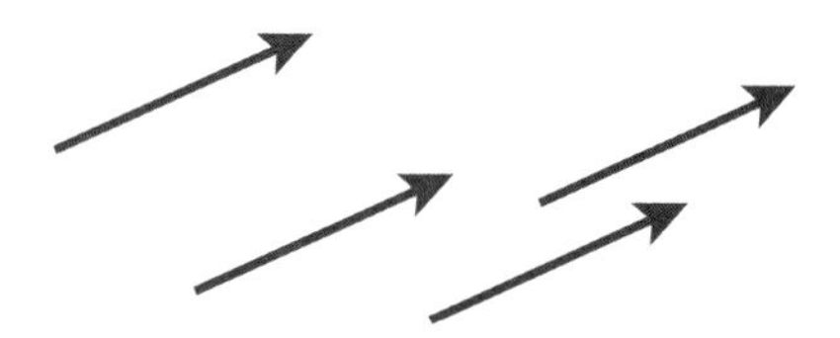

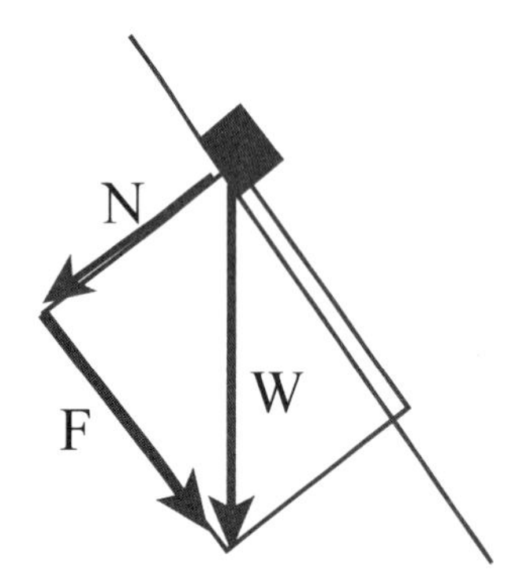

Now I'm going to move the F vector.

I moved it to the opposite side of the rectangle so its direction and length stayed the same.

Why did you do that?

Patience, my little man. You wanted me to explain how the slope $(= \frac{\text{Rise}}{\text{Run}})$ is equal to the coefficient of static friction $(\mu_s = \frac{F}{N})$.

(It seemed strange to Fred to be called, "my little man" by his doll who is only about six inches tall.)

Now with a little bit of geometry **(in the next chapter)** I will show that $\frac{F}{N}$ equals $\frac{\text{Rise}}{\text{Run}}$

Your Turn to Play

Trace each of these diagrams on a sheet of paper. (Do not write in this book!) Then do the artwork to resolve W into a normal force N and the frictional force F.

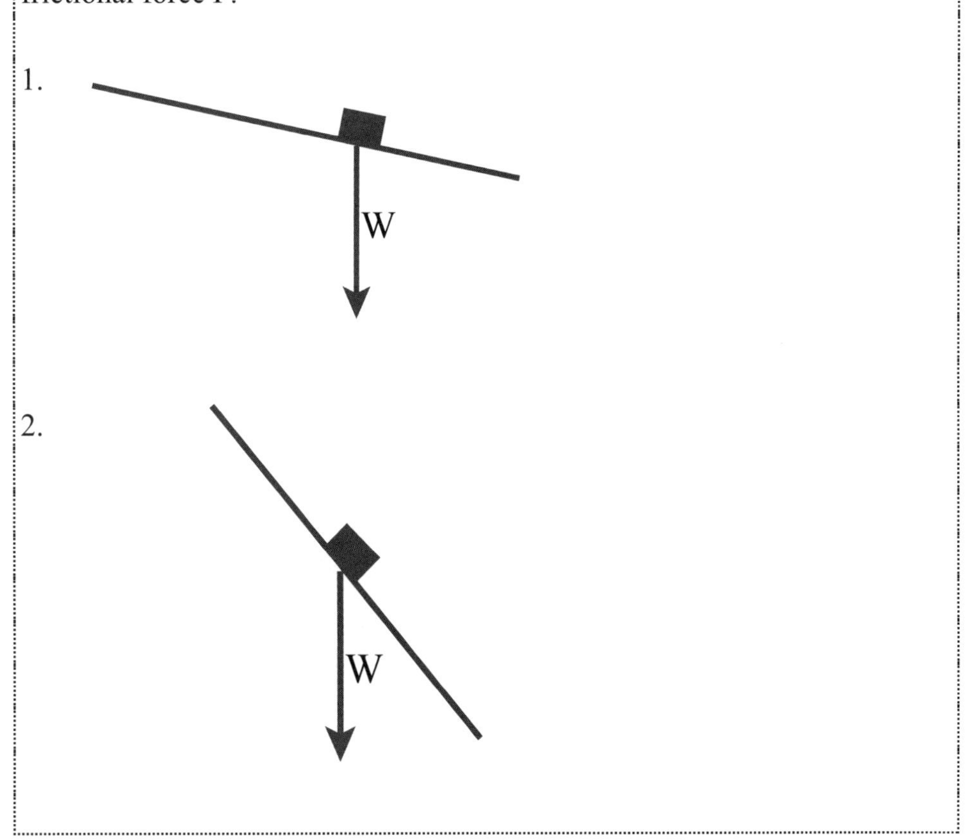

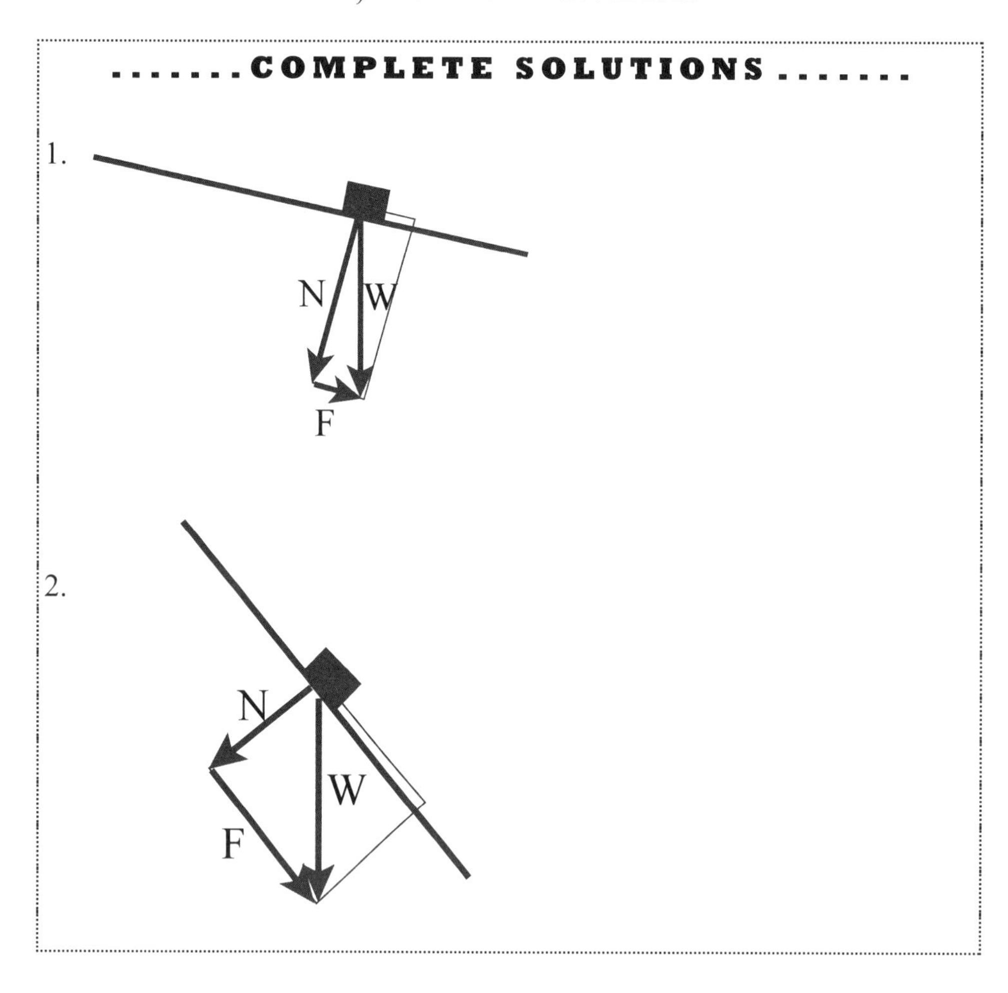

Some notes that probably aren't worth reading but will make things a bit more technically correct.

♪ #1: When we add two vectors, one way to do it is to move them so that the tail of one is at the head of the other. Then the sum of the vectors is from the tail of the first to the head of the second.

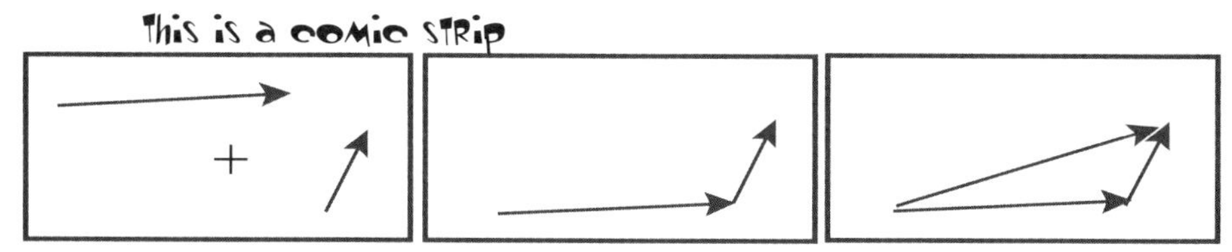

♪ #2: Addition of vectors is not adding their lengths. The length of the sum of two vectors is usually shorter than the sum of the lengths of the two vectors.

♪ #3: You can't divide vectors. (Don't ask me why.) When I wrote $\mu_s = \frac{F}{N}$, I meant that the length of F is divided by the length of N. Officially, this is written as $\mu_s = \frac{\|F\|}{\|N\|}$

Chapter Fifteen
Math Makes Things Easier

Resolving the weight force of the safe into the normal force and the force to overcome friction was new to Fred. It was physics. You put the toy safe on the surface. You tilt the surface until the safe starts to slide. Then you draw W. You draw N perpendicular to the surface. You make a rectangle. You get the lengths of F and N. You divide F by N and you get μ_s.

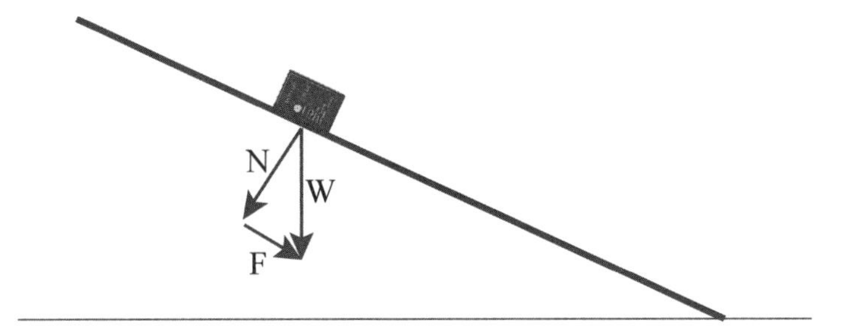

The big pain comes when you try to do **physics** in the **physic**al world. How do you easily measure the length of F and N when you have a book on a piano bench or a cookie on a convection oven door?

Kingie said *You don't have to crawl under the piano bench and make that drawing. You just measure the Rise and the Run of the surface when the object starts to slide.*

It's clear that $\frac{F}{N}$ *equals* $\frac{\text{Rise}}{\text{Run}}$

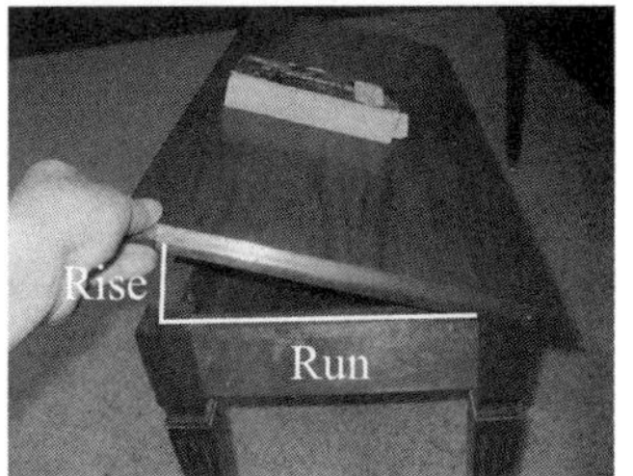

Wait a minute! I, your reader, object. It may be clear to Kingie. It may be clear to Fred. But it sure isn't clear to me! Are you are telling me that $\frac{F}{N}$ which is μ_s is somehow equal to $\frac{\text{Rise}}{\text{Run}}$?

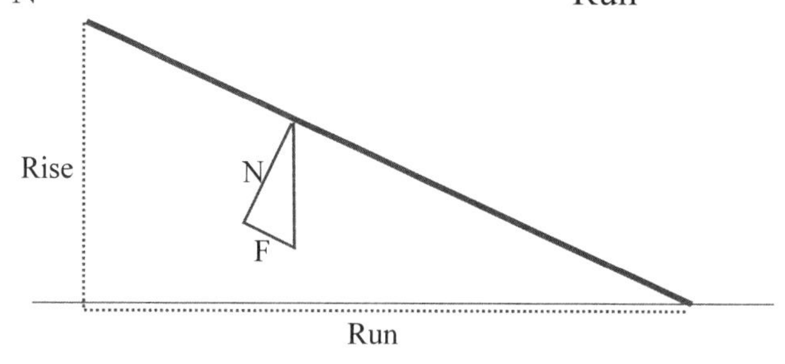

Oops. I forgot that you haven't had geometry yet. To begin this book you needed to know only fractions, decimals, and percents.

After this book comes

Life of Fred: Pre-Algebra 1 with Biology
Life of Fred: Pre-Algebra 2 with Economics
Life of Fred: Beginning Algebra Expanded Edition
Life of Fred: Advanced Algebra Expanded Edition

and then *Life of Fred: Geometry Expanded Edition.*

And on page 298 of that geometry book we begin the study of triangles that have the same shape.

You mean like the two triangles in the drawing above?

Exactly. The three angles in the little triangle match up with the three angles in the big triangle.

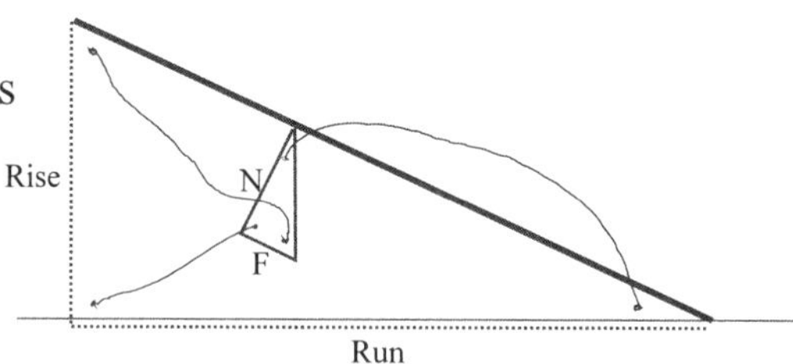

When that happens, the two triangles are called similar triangles.

If the two triangles are similar, the sides are proportional: $\frac{F}{N} = \frac{\text{Rise}}{\text{Run}}$

Here's the reasoning in a nutshell: If the three pairs of angles are equal, then the triangles are similar. If the triangles are similar, then $\frac{F}{N} = \frac{\text{Rise}}{\text{Run}}$

Similar triangles have the same shape. Everything stays in proportion. It is like enlarging a photograph. If Kingie is one-sixth as tall as Fred in one photograph, he will be one-sixth as tall in an enlarged picture.

Fred liked that photograph a lot. It made him look tall in comparison to a six-inch tall doll.

On the other hand, a photo of him standing next to a fire hydrant made him feel shorter.

Tiny facts from geometry:

#1 Every right angle is equal to 90° (ninety degrees).

#2 A little box indicates it is a right angle.

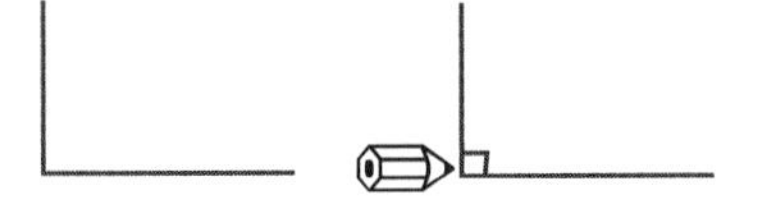

#3 The angles in any triangle always add up to 180°. (We prove that in geometry.)

Your Turn to Do Some Geometry

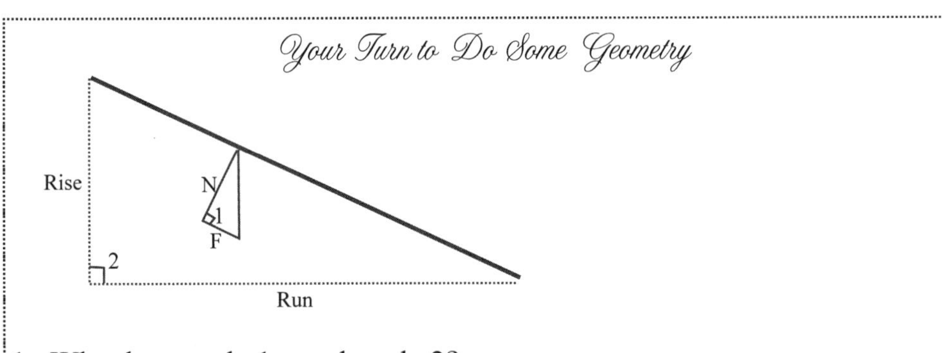

1. Why does angle 1 equal angle 2?

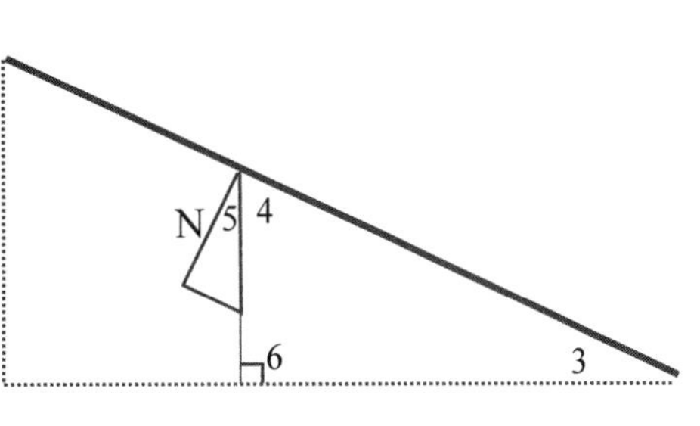

2. Suppose angle 3 was equal to 30°. What would angle 4 equal?

3. (Harder question) What would angle 5 equal? (Recall that the line marked N is normal (perpendicular) to the surface.

.......COMPLETE SOLUTIONS.......

1. Angles 1 and 2 are both right angles. They are both equal to 90°. Therefore, they are equal.

2. Angles 3, 4, and 6 all add up to 180°. (That was tiny geometry fact #3.)
We know that angle 3 is 30°
We know that angle 6 is 90° because it is a right angle.
Therefore, angle 4 must be 60°.

3. Angle 4 combined with angle 5 make a right angle because N is normal to the surface.
Angle 4 is 60° from the previous problem.
Angle 4 and angle 5 add to 90° since they make a right angle.
Therefore, angle 5 must be 30°.

We have matched up two angles in the small triangle with two angles in the large triangle.

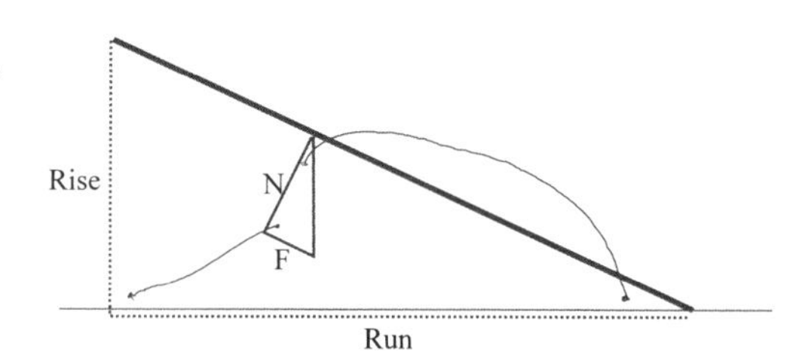

If the first pair of angles are equal and the second pair of angles are equal, then the third pair of angles must also be equal since the sum of the angles in any triangle is always 180°.

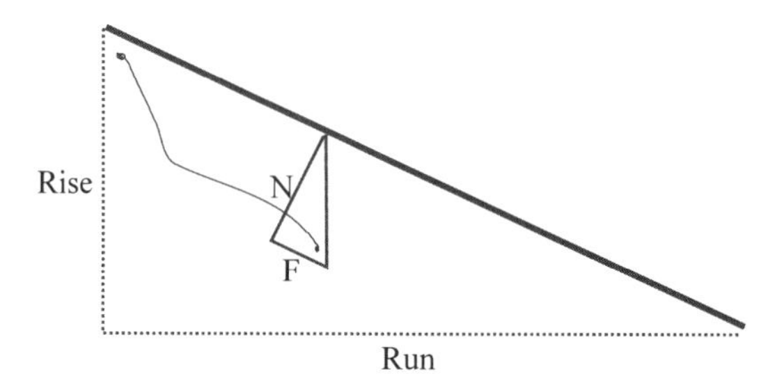

Chapter Sixteen
Work

Fred thanked Kingie for all the physics that he had explained.

He now understood:

❀ How to find the coefficient of static friction without knowing the weight of the object or the force needed to get it moving.

❀ How to resolve a vector, such as the weight vector W, into two component vectors N and F.

❀ How to add vectors such as N and F to get W.

❀ How the slope of the surface (= $\frac{\text{Rise}}{\text{Run}}$) is equal to μ_s (= $\frac{F}{N}$).

Fred already knew the mathematics (geometry):

✻ Every right angle is equal to 90º.

✻ The angles of any triangle will always add to 180º.

✻ Two triangles with the same shape are called similar triangles.

✻ Two triangles are similar if the three angles in one of them match up with the three angles in the other.

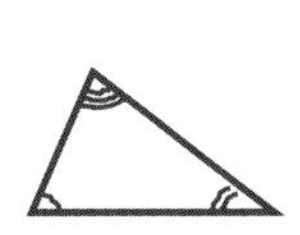

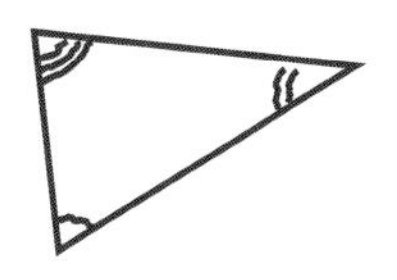

Kingie put on his painting hat and went back to working on another oil painting.

Fred tried to remember what he was doing before he was distracted by finding the safe in the hallway. He looked down.

He could see his knees.

He realized that he was in his jogging clothes.

He had been on his way for his morning jog.

When you get really old (like five years old), interruptions sometimes make you forget what you were doing.

Fred walked around the safe, over the door that lay on the floor, out into the hallway, past the nine vending machines (four on one side and five on the other), down the two flights of stairs, and out into the cool morning air.

He ran past the tennis courts, the university chapel, and the rose gardens. He came to the construction site of the newest building on campus—the Feynman Physics Building.

A giant crane lifted a 200-pound girder 40 feet into the air.

Fred computed in his head that the crane had done 8,000 ft-lb* of work.

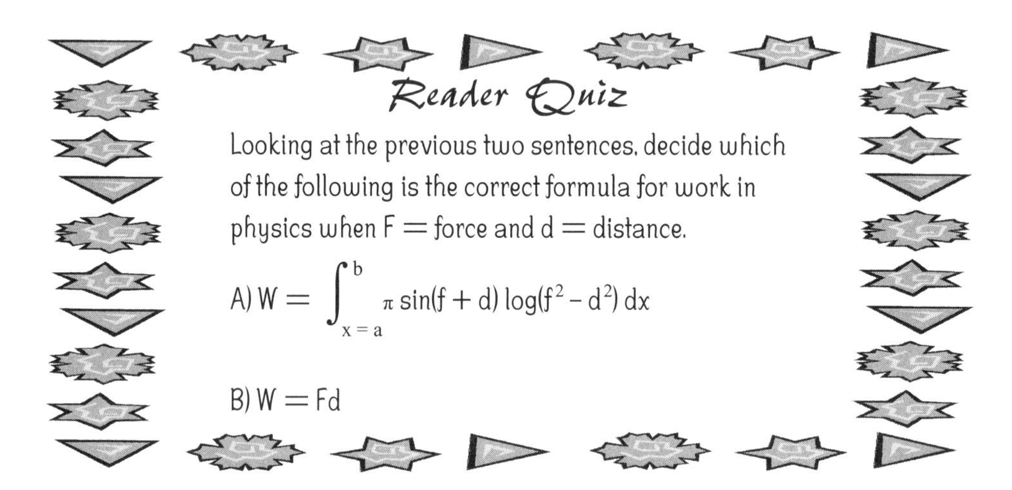

Reader Quiz

Looking at the previous two sentences, decide which of the following is the correct formula for work in physics when F = force and d = distance.

A) $W = \int_{x=a}^{b} \pi \sin(f + d) \log(f^2 - d^2)\, dx$

B) $W = Fd$

* Some physicists write foot-pounds as ft-lb or ft · lb or ft lb but none of them that I have run across write ft-lbs. or ft-lb. (using periods).

English can be so frighteningly difficult. It took me two minutes to compose the first sentence of this footnote. The first time I tried: *Some physicists write ft-lb but they never write ft-lb.* That looked dumb.

Then I wrote: *Some physicists write ft-lb but they never write ft-lb..* That made more sense (to me), but the style manuals prohibit two periods at the end of a sentence.

In contrast, lb. (with a period) is the correct way to abbreviate pound, and either lb. or lbs. (with periods) are the two correct ways to abbreviate pounds.

Answer to *Reader Quiz*: If you chose A), please select a different answer.

In physics the word *work* has a different meaning than *work* in everyday language. In physics work means force times distance. W = Fd.

Ask a physicist how much work is done in translating a hundred pages of German into Italian.

Ask a physicist how much work is involved in holding a door shut for two hours when a hurricane is trying to blow it open.

The answer in both cases is that no work has been done. In physics you have to *move* an object in order to perform work. Holding a door shut using 70 pounds of force is not work: W = Fd = (70)(0) = 0.

Your Turn to Play

1. *Work* is not the only word that has two different meanings. Name two different meanings for the word *degrees*.
2. Name two different meanings for the word *ounces*.
3. If you hand me your bowling ball, your suitcase, and your umbrella while you reach down and pick up a toothpick off the ground, which one of us is doing more work?
4. How much work is done in lifting a 37-pound child three feet up into your arms?
5. How many foot-pounds of work is done lifting an 8-ounce stapler 24 inches? (There are 16 ounces in a pound.)

.......COMPLETE SOLUTIONS.......

1. Degrees can be used to measure angles and they can be used to measure temperature. In addition, there are at least three different uses of degrees in temperature. 50° F is a cold room. 50° C is a very warm room. The Kelvin temperature scale is the same as the Celsius scale except that it is 273.15° colder. So 50° C = 323.15 K. (The degree symbol ° is not usually used in writing Kelvin temperatures.) The neat thing about the Kelvin temperature scale is that zero degrees Kelvin is **absolute zero**—the coldest possible temperature. Physicists have gotten to within a few millionths of a degree of 0 K.

90° angle

45° angle

2. Ounces can be a measure of weight. Sixteen ounces equals a pound.
Ounces can be a measure of volume. Sixteen ounces equals a pint.
The metric system doesn't have this confusion between weight and volume.

3. If you ask me, I would say that carrying all your stuff was a lot more work than picking up a toothpick. However, physicists say that without movement there can be no work. Picking up a toothpick and raising it four feet is considered work.

4. W = Fd = (37)(3) = 111 ft-lb

5. Since the answer is to be expressed in foot-pounds the first thing to do is change 8 ounces into pounds. We will use a conversion factor. We know that 16 ounces equals 1 pound, so the conversion factor will either be

$\frac{16 \text{ ounces}}{1 \text{ pound}}$ or it will be $\frac{1 \text{ pound}}{16 \text{ ounces}}$

We choose the one so that the ounces cancel:

$$8 \cancel{\text{ounces}} \times \frac{1 \text{ pound}}{16 \cancel{\text{ounces}}} = \frac{8 \text{ pounds}}{16} = \frac{1}{2} \text{ pound}$$

To change 24 inches into feet the conversion factor will either be

$\frac{12 \text{ inches}}{1 \text{ foot}}$ or it will be $\frac{1 \text{ foot}}{12 \text{ inches}}$

$$24 \cancel{\text{inches}} \times \frac{1 \text{ foot}}{12 \cancel{\text{inches}}} = 2 \text{ feet}$$

$$W = Fd = \frac{1}{2} \text{ pound} \times 2 \text{ feet} = 1 \text{ ft-lb}$$

Chapter Seventeen
Transfer of Energy

Watching the Feynman Physics Building being constructed was a lot of fun for Fred. He wandered around the construction site. (The construction workers had forgotten to put up a fence to keep the public from wandering onto the site.)

Fred stood right under the 200-pound girder that was 40 feet in the air.

He looked up and thought about that girder. *Work is a transfer of energy.** The crane had transferred energy from its motor (either chemical or electrical) to energy of motion (kinetic energy) and then to an energy of position (height). There was 8,000 ft-lb of energy now stored in that girder. If the hook broke and the girder fell, then that energy would first turn into kinetic energy (energy of motion).

After falling about 37 feet that kinetic energy would turn into two forms of energy:

Sound—splat!

Spring energy—as Fred got squished like a spring that is compressed. He would be compressed beyond the proportional limit (where F = kx remains a good approximation). Beyond the elastic limit (where he could spring back to his original shape after the girder was removed). He would be in the plastic region (where there would be permanent deformation of his little body).

"Hey kid! What are you doing?" one of the workmen yelled.

"I'm watching you construct the new building," Fred answered.

"This is not a place for little kids. Please leave the construction area. You might get hurt."

---

* Recall the nine forms of energy: chemical, electrical, heat, height, light, motion, nuclear, sound, spring. There may be other forms, but this is a good starting list.

"Can I stay if I wear a hard hat like the one you are wearing?"

"Out!"

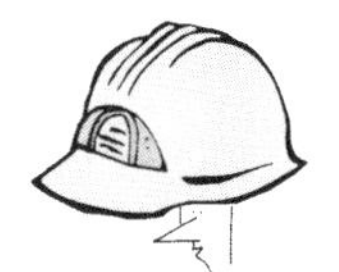
What Fred would look like in a hard hat

Fred was pretty smart in mathematics, but he didn't realize that a hard hat would offer little protection against a falling 200-pound girder.

Fred backed up. The workman put up a temporary orange plastic fence.

Fred daydreamed about what it would be like to be a construction worker when he grew up. He could operate the crane. He could wear a hard hat. He could yell at five-year-olds and tell them to leave the construction area.

It is one of the important tasks for kids to imagine doing various jobs as grown-ups.

He cleared an area in the dirt and declared that this was the Fred Gauss construction site. He put a big leaf on his head. That was going to be his hard hat. Then he picked up a twig and tied a string on it. This was going to be a pretend 200-pound girder being pulled by a pretend bulldozer.

He pulled on the string. It took 80 pounds of pretend force to move the girder (twig) at a constant speed over a pretend distance of 300 feet. His bulldozer (his hand) had performed 24,000 ft-lb of work.*

That's a lot of work. He wiped the pretend sweat off his forehead.

When he stopped pulling on the string, the twig stopped moving. He wondered *Where did all that energy go? Twenty-four thousand foot-pounds of energy can't just disappear.*

Fred was right. Energy can't ever disappear, and it can't be created. There are the same number of foot-pounds of energy in the universe today as there was when the sun first began shining.

This is called the **Law of Conservation of Energy.**

---

* $W = Fd = (80)(300) = 24,000$

The Law of Conservation of Energy states that in any closed system, the amount of energy cannot change.

The universe is a closed system.

The Fred Gauss construction site can be a closed system (as long as no energy is put into it from the outside—like the sun shining on it—and as long as no energy escapes from it).

A rocket ship in space and the space around it can be a closed system.

Since matter* is one form of energy,** it is correct to say:

The Universe is Pure Energy.

Your Turn to Play

1. As Fred pulled on the string, it took 80 pounds of pretend force to move the 200-lb. girder (twig) at a constant speed over a pretend distance of 300 feet.
Find the coefficient of kinetic friction.
2. Is it possible to find μ_s from the information in problem 1?
3. When Fred stopped pulling on the string, the twig stopped moving. Where did the kinetic energy (the 24,000 ft-lb) go? (Hint: It didn't turn into light or sound or electrical or height or nuclear or spring or chemical energy.)
4. If 80 pounds of pretend force would move the girder (twig) at a constant speed, what would happen if he applied 90 pounds of force?

---

* *Matter* is the physics word for stuff. Everything you can touch is matter. Matter can be atoms, Rag-A-Fluffy dolls, or a plate of spaghetti.

** That is what Einstein's $E = mc^2$ means.

.......COMPLETE SOLUTIONS.......

1. $F = \mu_k N$ $\frac{F}{N} = \mu_k$ $\frac{80}{200} = \mu_k$

Reducing $\frac{80}{200}$ by dividing top and bottom by 40, $\mu_k = \frac{2}{5}$ (or 0.4).

2. It is not possible. All we know is that the coefficient of static friction is greater than or equal to the coefficient of kinetic friction. $\mu_s \geq \mu_k$

3. Friction turns the energy of motion into heat.

4. Applying 90 pounds of force will cause the girder to move faster and faster. It will accelerate.

Perpetual Motion Machines

Pinwheels can spin in the wind. That doesn't violate the Law of Conservation of Energy since the wind is energy coming from the outside.

If we put the pinwheel in a box and close the lid, it will stop spinning.

A perpetual motion machine is a machine that will keep working inside of a closed box (a "closed system"). If we listen to it working, sound energy is coming out of the box . . . forever. That closed system is creating energy. If the Law of Conservation of Energy is true, perpetual motion machines cannot exist.

Here is my pinwheel perpetual motion machine—three balls equally spaced around the pivot point. Isn't it pretty? The two balls on the left outweigh the single ball on the right so it should spin counterclockwise and keep going if I give it a push.

But it won't.

Chapter Eighteen
Storing Energy

When Fred played in the dirt and dragged his pretend girder (twig) with his pretend bulldozer (his hand) he had used 24,000 foot-pounds of energy. And it had all turned into heat. The energy was lost. He couldn't get it back.

The Law of Conservation of Energy says that the energy in the closed system of the Fred Gauss construction site stays constant. But the energy (heat) was no longer in a form that was useable.

Then he took his pretend 200-pound girder and lifted it 40 feet into the air. $W = Fd = (200)(40)$ He used 8,000 foot-pounds of energy, but this time the energy was not lost. He could get it back. Just letting go of the girder will transform the height energy into motion (kinetic) energy.

If the girder landed on a giant spring, it could compress the spring and turn the kinetic energy into spring energy.

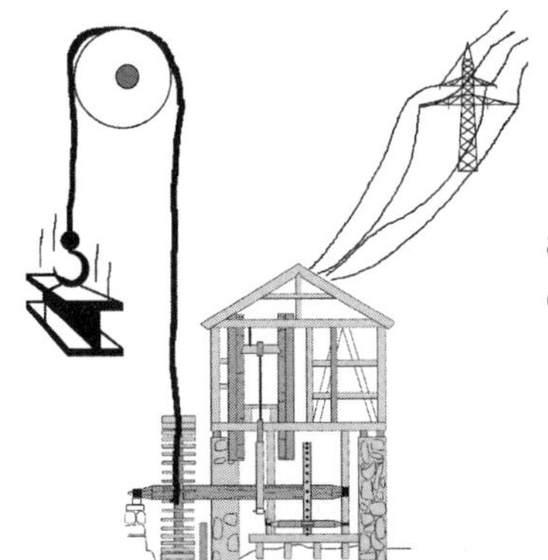

If the cable holding the girder were attached to a generator, the falling girder could spin the generator and create electricity.

Motors turn electricity into motion. Generators turn motion into electricity. Large generators supply the electricity to your house. Your electric bill is a charge for the amount of energy you have used. Instead of measuring the electrical energy in foot-pounds, it is measured in kilowatt-hours.

2,655,000 ft-lb = 1 kilowatt-hour

There are three common ways to store energy.

When Fred winds up his toy monkey, he stores energy in the spring.

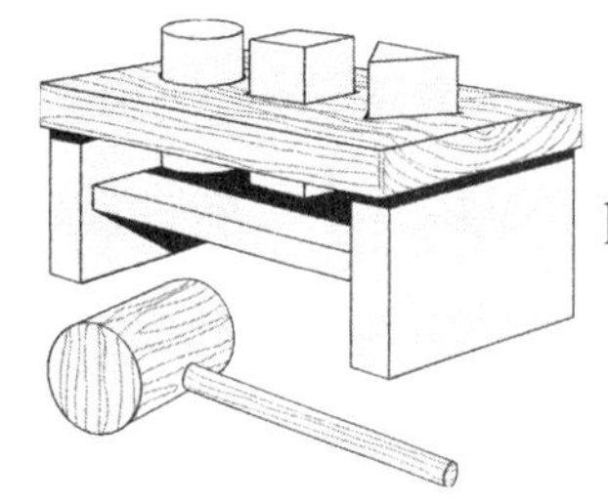

When he raises the hammer, he stores height energy.

When he grows strawberries in the windowsill of his office, he stores the light energy of the sun as chemical energy in the sweetness of the strawberries. This is called photosynthesis.

The energy in food is usually measured in Calories.
3,088 ft-lb = 1 Calorie

Energy can also be measured in ergs, British thermal units, joules, or electron volts.

Your Turn to Play

1. Convert 13,896 foot-pounds into Calories. Use a conversion factor. (3,088 ft-lb = 1 Calorie)

2. Here is a photo from a butter carton. One tablespoon (Tbsp.) of butter could lift a 200-pound man how far into the air?

3. We have to get back to my pinwheel perpetual motion machine to explain why doesn't spin.

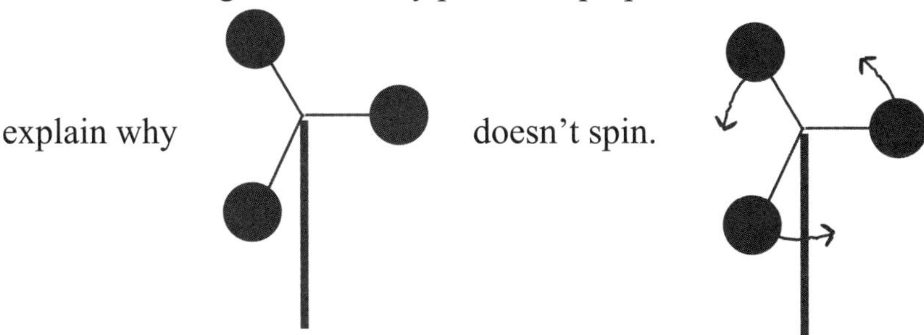

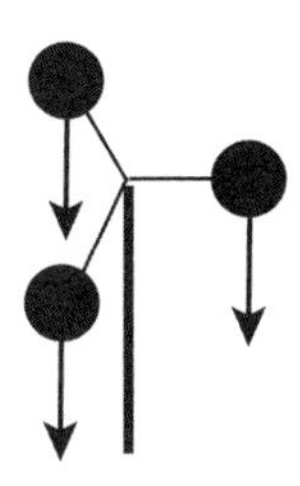

The three balls all weigh the same amount so the weight vectors for all three balls are alike. People who don't know physics might think that the weight of the two balls on the left would be more than the weight of the one on the right and would cause the pinwheel to spin.

But what causes it to spin is not the downward force, but the force perpendicular to each of the axes.

The ball on the right has all of its weight acting perpendicular to its axis.

The weight of the top ball is resolved into two forces: one perpendicular to its axis and one that is along the line of the axis.

It is time for your art. Copy this top ball on a sheet of paper and resolve the weight vector into the vectors perpendicular and parallel to its axis.

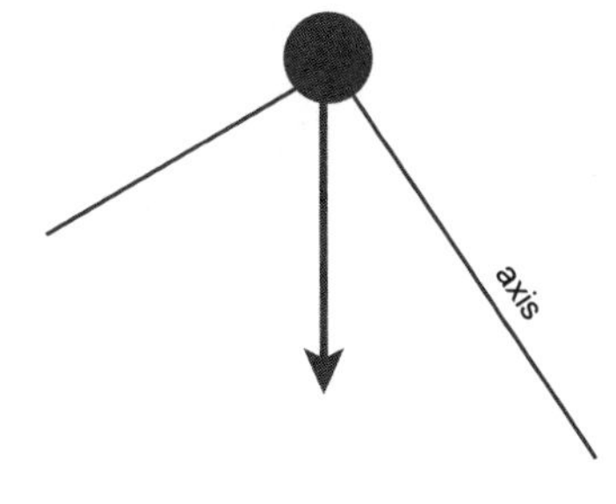

.......COMPLETE SOLUTIONS.......

1. Since 3,088 ft-lb = 1 Calorie, the conversion factor will either be

$$\frac{3088 \text{ ft-lb}}{1 \text{ Cal}} \text{ or it will be } \frac{1 \text{ Cal}}{3088 \text{ ft-lb}}$$

We chose the one so that the foot-pounds will cancel.

$$\frac{13896 \cancel{\text{ft-lb}}}{1} \times \frac{1 \text{ Cal}}{3088 \cancel{\text{ft-lb}}} = \frac{13896}{3088} \text{Cal} = 4.5 \text{ Calories}$$

$$\begin{array}{r} 4.5 \\ 3088 \overline{)\,13896.0} \\ \underline{12352} \\ 15440 \\ \underline{15440} \end{array}$$

2. According to the label, 1 tablespoon of butter has 100 Calories. Since one Calorie equals 3,088 foot-pounds, a hundred Calories equals 308,800 foot-pounds. (To multiply by a hundred you add a couple of zeros.)

We want to find a distance.

W = Fd

Divide both sides by F $\quad \frac{W}{F} = d$

We know that W = 308,800 ft-lb and that F = 200 lbs.

$$d = \frac{W}{F} = \frac{308800}{200} = 1544.$$

A tablespoon of butter could lift a 200-lb. man 1,544 feet into the air.

The Empire State Building is 1,250 feet tall. It takes a lot of exercise to work off a tablespoon of butter. Climbing its 102 stories works off about four-fifths of a tablespoon of butter.

3. Make the rectangle. Resolve the weight vector into its two components.

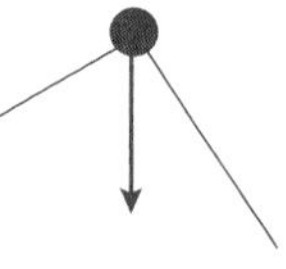

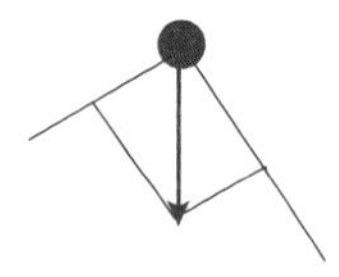

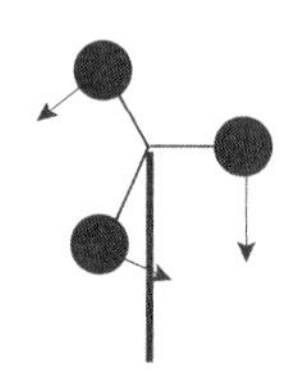

If we find the perpendicular vectors (the ones that will cause the pinwheel to spin) for all three balls, the lengths of the two on the left will exactly add to the one on the right. The pinwheel is balanced and will not want to spin.

The Bridge

from Chapters 1–18

first try

Goal: Get 9 or more right and you cross the bridge.

1. Joe pulled his 210-lb. boat with a force of 60 pounds along the sandy beach of the Great Lake. The coefficient of kinetic friction between the boat and the beach is 0.25. Will the speed of the boat remain constant? (Show your work. Just guessing either yes or no is not enough.)

2. If Joe pulled his 210-lb. boat 100 feet at a constant speed along the beach (where $\mu_k = 0.25$), how much work would he do?

3. It took 63 lbs. of force to get Joe's 210-lb. boat moving. What is the coefficient of static friction? Express your answer as a decimal.

4. On Joe's first fishing trip this month, he used 5 lbs. of bait and caught 3 fish. On his second trip, he used 8 lbs. of bait and caught 4 fish. On his last trip, he used 2 lbs. of bait and caught 1 fish. Plot these three points where the first coordinate is the weight of the bait he used and the second coordinate is the number of fish he caught.

5. Because Joe just eats candy bars while he is fishing, he is gaining $3\frac{1}{3}$ pounds each month. How long will it take him to gain 25 pounds?

6. Joe likes 6 tablespoons of melted butter on his popcorn. Using conversion factors, determine how many foot-pounds of energy are in those 6 tablespoons. (3,088 ft-lb = 1 Calorie) No credit for this problem if you don't use a conversion factor. First convert tablespoons of butter to number of Calories, then Calories to foot-pounds.

Nutrition Label for Butter

7. If 36y = 75, then what does y equal?

8. If two angles of a triangle measure 49° and 78°, what is the measure of the third angle?

9. A fishhook costs 30¢. Joe figured out that two fishhooks cost 60¢. Is this an example of inductive or deductive reasoning?

10. As Joe watched the sun set over the Great Lake, he thought about c = 299,792,458 meters per second. Is c = 299,792,458 m/s used to establish the speed of light or the length of a meter?

The Bridge

from Chapters 1–18

second try

---

1. Darlene was planning to have two bridesmaids at her wedding until she read in *Royal Weddings* magazine that Princess Lucre (pronounced LEW-ker) had 17 bridesmaids. Is the number of bridesmaids a discrete or continuous variable?

2. The wedding dinner of Princess Lucre cost $5,000 for each table of guests. There were 8 at each table. Using a conversion factor, determine how much 20 tables of guests would cost.

3. At the wedding dinner, Queen Boodle (Lucre's mom) sat on a large chair. Boodle and the chair weighed a total of 300 pounds. After she sat down, a servant pushed her and the chair up to the table. It took a force of 180 pounds to get the queen and her chair moving. Find the coefficient of static friction between the chair and the floor. Express your answer as a decimal.

4. Darlene thought it would be nice to sit down at her wedding and have Joe shove her toward the table. The coefficient of static friction in this case is 0.8 since Darlene's wedding dinner would be held at her apartment, which has carpet instead of a polished floor like Queen Boodle's palace. It would take 120 pounds of force for Joe to get Darlene and her chair moving. How much do Darlene and her chair weigh together?

5. When Queen Boodle read the first 80 names on the guest list that her daughter had created, she noticed that they were former boyfriends of Lucre. She reasoned that all 432 names on the guest list would be former boyfriends. Was this an example of deductive or inductive reasoning?

6. *Royal Weddings* had an article on what Princess Lucre did on the night before her wedding. The first thing she did was have a bath in her giant tub. "I feel filthy," she said. The tub holds 200 gallons of water and fills at the rate of 40 gallons per minute. How long would it take to fill the tub?

The Bridge

from Chapters 1–18

7. *Royal Weddings* reported that while Lucre's bath was being drawn (the tub was filling), she played with a bar of soap. She put it on a toilet seat lid and raised the lid until the bar started to slide.

Here is the picture that was in the magazine.

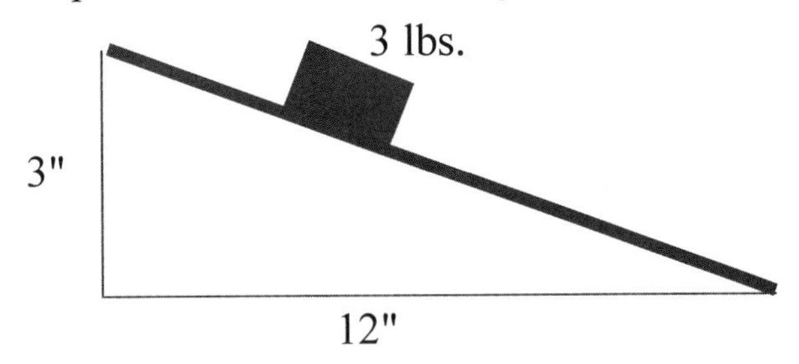

Find μ_s between the bar of soap and the toilet seat lid.

8. Is it possible to find μ_k from the information in the *Royal Weddings* picture?

9. Copy that *Royal Weddings* picture on a sheet of paper and draw the weight of the bar of soap vector arrow. Label that vector W.

10. Continue the drawing you made in the previous problem by resolving W into the normal force N and the force F which is parallel to the surface.

(Lucre and boodle are words in the dictionary.)

The Bridge

from Chapters 1–18

third try

---

1. Fred liked reading Christina Rossetti's poetry. In reading some of her melancholy poems, Fred would shed 3 tears for every 8 lines that he read. Using a conversion factor, determine how many tears Fred would shed reading 112 lines. (No credit for this problem if you don't show your work and use a conversion factor.)

2. In her poem "Memento Mori" Fred read the lines:

Sweet the sorrow
Which ends tomorrow.

He thought that memento mori had something to do with getting over a cold. (It doesn't.) He knew that the sorrow of having a cold is bearable since there is a tomorrow when he would be well again.

When Fred caught a cold last year, Kingie painted him a get-well painting. It measured 16 inches by 20 inches. What was the area of that painting?

"Tomorrow"
by Kingie

3. It takes 0.3 pounds of force to slide a copy of Christina Rossetti's poetry across the desk at a constant speed. The coefficient of sliding friction of her book and the desk is 0.6. How much does the book weigh?

4. Fred usually read Rossetti's work at the rate of 12 pages per hour. That isn't a very fast reading rate, but Fred spent a lot of time memorizing as he read. If you said to him, "Sweet the sorrow," he would respond, "Which ends tomorrow." How many hours would it take him to read 60 pages?

5. Fred put a dime on the cover of his Rossetti poetry book. Is it possible to find μ_s for the dime and the cover of the book without knowing the weight of the dime?

6. One of the most often used formulas in algebra is d = rt, where d is distance, r is rate, and t is time. Find t in terms of d and r.

The Bridge
from Chapters 1–18

7. Having a cold is a big event in Fred's life given the size of his nose. One cold lasted 5 days and he used 3 boxes of tissue. Another cold lasted 7 days and he used 5 boxes. Plot these two points where the number of days is the first coordinate.

8. Here is a picture of the tissue box. Which is the larger numeral?

9. Which is the larger number on that box?

10. Which is the largest digit on that box?

The Bridge

from Chapters 1–18

fourth try

1. Jake, Jack, and Jane were three sailors visiting New York City. Jack was shopping for electronics of all kinds—radios, cell phones, vacuum tubes, etc. He started with an empty duffle bag. It was getting $2\frac{1}{2}$ pounds heavier each hour with the things he was buying.

How long would it be before Jack had $11\frac{1}{4}$ pounds of electronics in his duffle bag?

2. Jack's duffle bag weighed 160 pounds. It took 64 pounds of force to drag it on the sidewalk. What is the coefficient of kinetic friction between his duffle bag and the sidewalk?

3. The three sailors came to a large triangular park. How many degrees are in the third angle in this triangle?

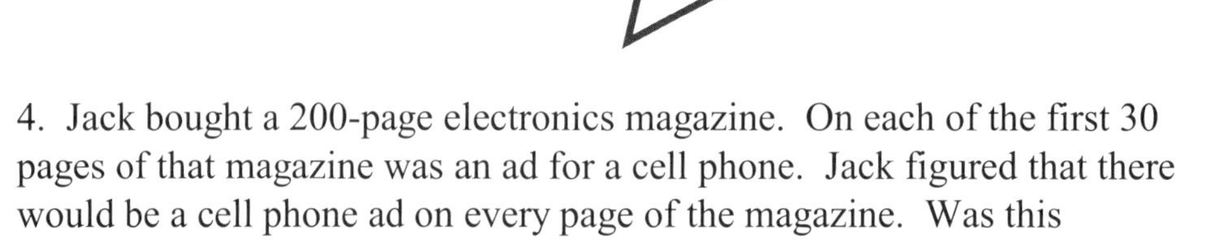

4. Jack bought a 200-page electronics magazine. On each of the first 30 pages of that magazine was an ad for a cell phone. Jack figured that there would be a cell phone ad on every page of the magazine. Was this inductive or deductive reasoning?

5. In the park, Jane decided to play on the slide. It wasn't very slippery. In fact, when she got on, she didn't slide at all.

Jake and Jack fixed that. They lifted up the back end of the slide until Jane started to slide. Find μ_s.

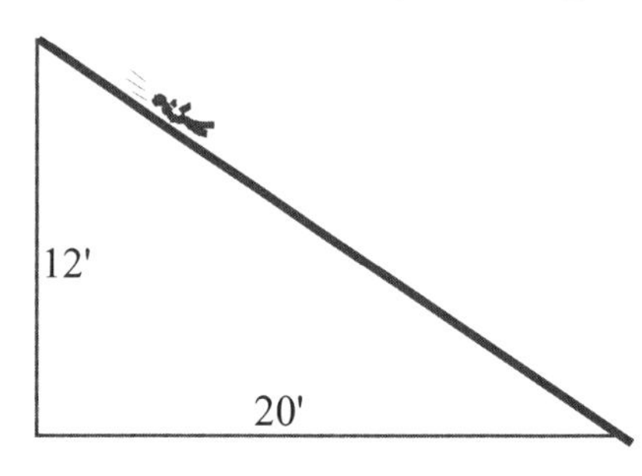

The Bridge

from Chapters 1–18

6. Jack was buying lots of electronic gear. Every 24 minutes he would spend \$16. Using a conversion factor, determine how long it would take him to spend \$56. (No credit for this problem unless you use a conversion factor.)

7. If he spent \$16 every 24 minutes, how much would he spend in 90 minutes? Use a conversion factor.

8. If $8x = 3$, what does x equal?

9. Is the number of pigeons in the triangular park a discrete or a continuous variable?

10. Fill in the blanks.

A motor changes ____?____ energy into _____?_____ energy.

The Bridge

from Chapters 1–18

fifth try

1. PieOne is noted for unusual pizzas. Their personal size pizza will feed three people. Their regular size pineapple and pepperoni pizza weighs 27 pounds. How much work is involved in lifting that pizza three feet upward?

2. Stanthony would only put pizzas in the oven when it had reached the perfect temperature. While the oven was heating, he held the 27-pound pizza steady in his arms. He didn't want the pineapple and the pepperoni to wiggle around. How much work did he do holding that pizza for ten minutes while the oven heated?

3. When the oven was at the perfect temperature, he slid the 27-pound pizza into it at a constant speed. If μ_k between the pizza and the oven is 0.4, how much force did Stanthony need to apply?

4. He could make 18 pizzas in 27 minutes. Using a conversion factor, determine how long it would take to make 24 pizzas.

5. With the help of the customers, Stanthony carried a large clam and calamari pizza to their table. This Sea Captain's® pizza weighed 77 pounds. After they put it on the table, they realized that they needed to give it a push to move it to the center of the table. With a shove of 11 pounds they could get it moving. Find μ_s. Express your answer as a fraction.

6. To give the Sea Captain's® pizza extra flavor, Stanthony would pour starfish oil over the top before baking it. Is the volume of starfish oil a discrete or continuous variable?

7. If 6x = 12, what does x equal?

8. What is the name of the law which states: "In any closed system, the amount of energy cannot change"?

9. Fill in the blank: Friction changes the energy of motion into ____?____.

10. Plot (2, 5), (3, 5), and (4, 2).

Starfish
Oil

Chapter Nineteen
Metric System

Fred was getting tired of playing in the dirt, dragging his pretend girder (twig) with a pretend force of 80 pounds over a pretend distance of 300 feet, doing 24,000 foot-pounds of pretend work.

Wait a minute. I, your reader, am getting tired of all these Imperial (British) units of feet and pounds. Most of the world does things in the metric system. Why aren't you using kilograms and meters?

You probably would not like it if I switched over to the metric system.

Everything about the metric system is wonderful. You have a thousand grams in a kilogram. You have a thousand meters in a kilometer. You don't have a silly twelve inches in a foot and 5,280 feet in a mile.

If Fred were playing in the dirt in the metric system, you would probably go nuts.

Try me! I dare you!

Okay. Fred was playing in the dirt dragging his pretend girder (twig) with a pretend force of 80 newtons over a pretend distance of 300 meters, doing 24,000 joules* of work.

Hey! Where's the kilogram? Where's the kilogram-meter? You have got a lot of explaining to do, Mr. Author. I want my kilogram and I want it now!

Okay. How much do you weigh?

That's none of your business! I know I'm a little heavier than Fred who only weighs 37 pounds. Let's talk about Fred's weight. Besides, the numbers will be smaller and easier to work with. I know that a kilogram is a little more than two pounds.

How about 2.2 pounds?

I'll accept that.

---

* Joules is pronounced JEW-els. James Prescott Joule lived in the 1800s. He invented the foot-pound as the unit of work. I have no idea how his name got attached to the metric unit (joules). This would be the same as if you worked with donkeys and they named a car after you.

Will you accept 1 kg = 2.205 lbs.? Or better yet, 1 kilogram = 2.20462262 pounds.

Okay. What's your point? We just use a conversion factor and turn his 37 pounds into kilograms. $\frac{37 \text{ lbs.}}{1} \times \frac{1 \text{ kg}}{2.20462262 \text{ lbs.}}$

My point is that 1 kilogram doesn't equal 2.20466262 pounds.

Big deal. It is approximately equal to that. I remember back in Chapter 4 where you showed that light travels in a vacuum exactly 299,792,458 meters in a second. I understand that in this case the conversion from kilograms to pounds will not be exact.

It is worse than that. Please sit down. I need to tell you the facts of life that your mother probably never told you.

You can't convert feet into hours.

You can't convert gallons into square inches.

You can't convert pounds into kilograms.

Feet is a measure of length.

Hours is a measure of time.

Gallons is a measure of volume.

Square inches is a measure of area.

Pounds is a measure of force. When Fred steps on a scale, the scale says 37 pounds of force are being applied.

On the moon, gravity is about one-sixth of what it is on earth. If Fred stepped on a scale on the moon, the scale would read 6 pounds. (One-sixth of 37 pounds is about 6 pounds.)

Kilograms is not a measure of force. It is a measure of mass. Mass is a measure of how much matter something has. Matter is the official name for "stuff."

If Fred has a mass of 17 kg when he is playing in the dirt on the KITTENS campus, he will have a mass of 17 kg if he were playing in the dust on the moon. The number of atoms in his body hasn't changed, so his mass hasn't changed.

Wait a minute! My friend in Germany* emailed me last week that she went on a diet and now weighs 55 kilograms. Is she wrong?

No. She is a normal human being. It is physicists who talk funny. Physicists will say that it is more work to pick up a paperclip than it is to hold six bowling balls in your arms for an hour. Work = Force × distance.

Physicists will say that your German friend has a mass of 55 kilograms, but her weight depends on whether she is standing on Earth or standing on Jupiter. In outer space she would have no weight at all.

Okay. This is probably an easy question. How do you measure the mass of something? Do you have to go and count the atoms?

That would be the hard way. Oops. We have to leave room for the *Your Turn to Play*. Ask me at the beginning of the next chapter.

Your Turn to Play

1. Some things are exactly equal. There are 12 inches in a foot. Light travels at exactly 299,792,458 meters per second in a vacuum. One meter is exactly 100 centimeters. They are exact because we *define* a foot to be exactly 12 inches. We *define* a meter to be the distance light travels in 1/299,792,458 of a second.

Some things are approximate. π ≈ 3.1415926535897932384626433. One liter ≈ 1.057 quarts. (≈ means "approximately equal to.")

Which of these are exactly equal?

A) one dozen ↔ twelve

B) one year ↔ 365 days

C) one hour ↔ 60 minutes

D) Fred's weight ↔ 37 pounds

2. In the Imperial (British) system, we can measure energy (or work) in foot-pounds. In the metric system, we can use joules. One foot-pound is approximately equal to 1.4 joules. Using a conversion factor, convert 70 foot-pounds into joules.

* Or France or Italy or Russia or Japan or China or Peru or Egypt or anywhere the metric system is used.

.......COMPLETE SOLUTIONS.......

1. A) One dozen exactly equals 12 by definition.

B) One year does not equal 365 days. If it did, then calendar making would be a lot easier.

Some teachers tell their students that there are 365¼ days in a year. They say that this is the reason that we put an extra day in February every four years (a leap year).

A year is defined as the time it takes for the earth to make one trip around the sun. It doesn't take 365¼ days. One year is approximately equal to 365.242374 days.

The calendar that we use is called the Gregorian calendar. To make 365¼ days closer to 365.242374 days, years that end in 00 (those divisible by 100) are *not* leap years. Using that scheme, the average number of days in a year is $365 + \frac{1}{4} - \frac{1}{100} = 365 + 0.25 - 0.01$ which equals 365.24.

But 365.24 days isn't close enough to 365.242374 days. So the Gregorian calendar makes one more adjustment. 1700, 1800, and 1900 are non-leap years, but 2000 is a leap year. Those years evenly divisible by 400 are leap years. The average number of days in the Gregorian calendar year is $365 + \frac{1}{4} - \frac{1}{100} + \frac{1}{400} = 365 + 0.25 - 0.01 + 0.025 = 365.2425$ which is pretty close to 365.242374 days.

About once every 18 months, the International Earth Rotation and Reference Systems Service adds or subtracts a leap second.

C) One hour is exactly 60 minutes by definition.

D) Fred's weight varies all the time. If he takes a sip of water, he will gain weight. If he sweats, he loses weight.

2. $\frac{70 \text{ foot-pounds}}{1} \times \frac{1.4 \text{ joules}}{1 \text{ foot-pound}} = 98 \text{ joules}$

Chapter Twenty
Measuring Mass

Fred got tired of playing in the dirt and stood up. His mass had increased by a little bit since his clothes were now dirty.

Wait a minute! You told me you were going to tell me how to measure Fred's mass without having to count the atoms. You didn't keep your promise.

Yes I did. I said that I would tell you if you reminded me.

I'm reminding you!

Thank you. There are two different ways I can think of to measure the amount of matter Fred has.

The first way is to put Fred on a bathroom scale. That scale will measure his weight in pounds. It will measure the force with which he presses against the scale because of gravity.

The more mass an object has, the more weight it has on earth. Suppose football players have a mass of 100 kg each. Two football players would crush Fred twice as much as one football player.

Two football players would weigh twice as much as one player.

On earth, the mass of an object is proportional to its weight. So to measure the mass of two different objects on earth, all you need to do is measure their comparative weights.

On earth, if you put a one-kilogram object on a scale, the scale will read 2.2 pounds of weight. So a two-kilogram mass will weigh 4.4 pounds.

The scales made in Germany, France, Italy, Russia, Japan, China, Peru, or Egypt say "kilograms," but they are secretly measuring pounds and dividing by 2.2. Those scales should have a warning: GOOD ONLY FOR USE ON THE EARTH.

Fred has a mass of about 17 kg. If he used one of those German scales on the moon, it would read about 3 kg. (The gravity on the moon is about one-sixth of the earth's gravity.) But Fred's mass is 17 kg no matter where he is. The amount of matter in his body does not change.

Okay. But what is a kilogram? You are telling me that I can only use a scale to measure mass (kilograms) if I first know what the weight of one kilogram is.

That's easy. You go to the International Bureau of Weights and Measures, which is located near Paris. They keep a cylinder of solid metal under lock and key. Everyone agrees that that cylinder is one kilogram.

Official Kilogram

I remember that International Bureau of Weights and Measures. That's where before 1960 they used to define a meter as the distance between two lines scratched on a metal bar. And then in 1983, they finally defined a meter as the distance light travels in a vacuum in 1/299,792,458 of a second. They could throw away that metal bar if they wanted to or sell it on eBay.

So when are they going to throw away that one-kilogram cylinder of metal?

Just as soon as they can figure out how to substitute something better.

The goal is simple. Suppose a friend of ours lives on Mars. We can telephone him and describe exactly how long a meter is. We don't have to FedEx him a metal bar with two scratches on it. He just has to measure the speed of light.

But—this is embarrassing—we haven't yet figured out how to tell him what a one-kilogram mass is. There is no way that he can go to his lab and figure out what one kilogram is by anything we tell him on the telephone.

We can tell him length. 1 m = 1/299,792,458 of distance light travels in one second.
We can tell him temperature. 0°C = water freezes. 100°C water boils.
We can tell him the value of π. Take any circle and divide the circumference by the diameter.

We can tell him time. **Measure vibrations of a particular atom.**[*]

But if he wants to know what a kilogram of mass is, he has to travel to Paris, France, and examine this chunk of metal.

Someday we may have a definition of a kilogram that can be described over the telephone.

Your Turn to Play

1. (Question for art majors.)The circumference of a circle is the distance around the perimeter.

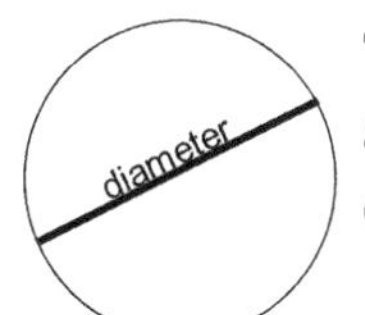

The diameter of a circle is the distance across a circle which passes through the center.

Draw a big circle on a sheet of paper. If you own a compass, use it. Otherwise put a small plate on the paper and trace out a circle.

Measure the circumference. This is not easy. Either move a ruler along the edge somehow or place a string on the circle and then straighten it out. Then measure the diameter. Divide the circumference by the diameter. The answer is called π.

2. Near the beginning of this chapter, I said that there are two ways to measure the amount of matter (the mass) that Fred has. In a sentence or two, can you briefly describe those two ways?

---

[*] To be more precise, the 1967 the Thirteenth General Conference on Weights and Measures defined the second of atomic time in the metric system as the duration of 9,192,631,770 periods of the radiation corresponding to the transition between the two hyperfine levels of the ground state of the caesium-133 atom.

.......COMPLETE SOLUTIONS.......

1. It doesn't matter how big or small the circle you drew is. The answer will always be the same. This can be considered a physics experiment. You are looking at nature to find out what is true. (When you were ten months old, you were performing physics experiments by sticking everything you could get a hold of—fingers, crayons, the cat's tail—in your mouth to see how it tasted.)

When you divided circumference by diameter, you should have gotten an answer of about 3. $\pi \approx 3$. That is apparently the answer in the Old Testament (I Kings 7:23).

2. The first way to find the mass of an object such as Fred is to stick him on a bathroom scale and compare his weight with the weight of a one-kilogram object. On earth, one kilogram weighs approximately 2.205 pounds. $\frac{\text{one Fred (37 lbs.)}}{1} \times \frac{1 \text{ kg}}{2.205 \text{ lbs.}}$ will do a nice job of converting pounds to kilograms on earth.

The second way to find the mass of an object is. . . .

I'm waiting. I, your reader, couldn't find your description of a second way to find the mass of an object. I looked over the chapter twice. You got distracted by talking about how we define one kilogram.

Oops. I guess I did. How can I make it up to you?

There is always Chapter 21. Am I going to have to remind you once we get to the next chapter?

Maybe if I entitle Chapter 21 as "A Second Way to Measure Mass," I might remember.

I hope so.

Chapter Twenty-one
A Second Way to Measure Mass

Fred jogged to the Great Lawn on the KITTENS campus. There was no one on the lawn. Fred took off his shoes and socks and ran barefoot across the lawn.

What Fred didn't notice was a woman standing on the edge of the Great Lawn practicing hitting golf balls.

The second way to measure mass is by measuring a thing called inertia (in-UR-sha, where UR rhymes with HER). Inertia is the tendency of an object to keep going in the same direction at the same speed.

As the golf ball bounced off Fred's forehead, its direction changed. Fred had been vertical. He became horizontal.

If she had been practicing with Ping-Pong balls instead of golf balls, Fred would have remained vertical (and conscious). If she had been practicing with billiard balls, Fred might be more than just unconscious (dead).

If a Ping-Pong ball, a golf ball, and a billiard ball are all traveling at the same speed, it is much easier to change the speed or direction of the ball with the least mass. Inertia is a second way to measure mass.

Is it easier to stop a train going 10 miles per hour or a kid on a tricycle going ten miles per hour?

Newton's Second Law says that if you want to produce some change in direction or speed, the force needed is proportional to the mass. If the train has a mass that is a million times more than the kid on a tricycle, it will take a million times more force to make the same change in speed.

When Fred changed the direction of the golf ball (by bouncing it off his forehead), it took seventeen times as much force as it would have taken to change the direction of a Ping-Pong ball that had 1/17 of the mass of a golf ball. (Golf balls = 45.9 grams. Ping-Pong balls = 2.7 grams.)

Some notes:

♪#1: Inertia is the force needed to change an object's speed or direction.

♪#2: Inertia is proportional to mass. For example, if you double something's mass, you double the amount of force needed to bring it to a stop.

♪#3: Not everything is proportional.

Example A: Getting hit by a Ping-Pong ball (which weighs one-seventeenth of a golf ball) does not produce one-seventeenth of the damage. If you hit Fred's forehead with 17 Ping-Pong balls, it might start to itch. In contrast, one golf ball knocked Fred out and gave him a big lump on his forehead.

Example B: It is pretty easy to jump over a one-foot fence without touching it. It is not ten times harder to jump over a ten-foot fence without touching it. It is a zillion times harder.

Example C: Want to learn German or French or Italian? Live in Germany, France, or Italy for a year and you will be able to speak and write (and think) in that language. But no matter how hard you try, you can't learn the language over the weekend.

Example D: Staying awake for 20 hours isn't that hard to do. Staying awake for 80 hours isn't four times as hard. It's torture.

♪#4: Isaac Newton invented Newton's Second Law (force is proportional to mass) in the early 1700s. Lots of things were happening back then. In 1714 George arrived in London and became king of England, but he didn't speak English! In 1720 wallpaper became fashionable in England. Also in 1720 some of your ancestors were alive.*

Your Turn to Play

1. In the really old days in Greece (say 2,500 years ago), they thought that the natural state of an object is to be at rest. Slide a penny across a table and it comes to a stop.

In the early 1700s Newton also wrote Newton's First Law: Every object continues to move at the same speed and in a straight line unless some outside force is applied to it.

The Greeks would have thought that Newton was nuts. They would have told Newton, "No one is stopping the penny. It just wants to stop because being at rest is the natural state for all objects."

What invisible force stops the penny?

2. Suppose the big test on French verbs is on Wednesday. On Tuesday at 7 p.m. you start studying. You learn 20 verbs in an hour. Will you know 100 verbs if you study until midnight?

---

* Would it be possible that none of your ancestors were alive in 1720?

.......COMPLETE SOLUTIONS.......

1. The Law of Conservation of Energy states that in any closed system, the amount of energy cannot change.

The penny and the table together create a closed system. After you push the penny, there are no external forces (outside the table and the penny) that affect the motion of the penny. There is no wind, no lightning strikes, and nobody reaching down and stopping the penny.

The penny has kinetic (motion) energy. Friction stops the penny by turning kinetic motion into heat. The table and the penny get slightly warmer.

If there were no friction, that penny would slide and slide and slide.

2. This is called cramming before an exam.

If you can learn 20 verbs in the first hour, does that mean you can learn 100 verbs in five hours?

The answer is yes—if you are a machine.

If you happen to be human, fatigue starts to set in and your ability to learn slows down. After about the third hour, your head is swimming and you can't remember the difference between *avaler* (to swallow) and *attrister* (to sadden).

During the first hour of study

After three hours of study

In the fifth hour of study

Chapter Twenty-two
Pressure

After the woman had finished hitting the 60 golf balls that were in her bucket, she put down the golf club, picked up the bucket, and went to collect the balls on the Great Lawn. She was unaware that one of her balls had hit Fred.

She found 59 golf balls and one little boy with a big bump on his head.

Fred sat up and asked, "What happened?"

"I think I hit you with one of my golf balls. I'm sorry."

"It must have been hit pretty hard," Fred said. "It knocked my shoes and socks off." Fred had forgotten that he had taken off his shoes and socks so that he could run barefoot across the lawn.

"It must have knocked my glasses off. Everything is blurry." Fred had forgotten that he didn't wear glasses.

She said, "Maybe you should wait a moment before you stand up."

Fred took her advice.

Soon things were much better. He stood up and introduced himself, "Hi. I'm Fred Gauss. I'm five years old and I live at KITTENS University.

"I'm Katherine Woods, but everyone calls me Kitty. I hit 60 golf balls, but I can only find 59 of them."

Fred thought 60 minus 59 equals one, and said, "I think you are missing one golf ball." He was glad that his injury had not knocked any mathematics out of his head.

He got out a piece of paper and checked his work.

$$\begin{array}{r} \overset{5}{\not{6}}\,{}^{1}0 \\ -59 \\ \hline 1 \end{array}$$

"The errant* golf ball might have gone into the Great Lake," she said. The Great Lake was right next to the Great Lawn.

When they got to the beach, they could see the track of the golf ball in the sand. It was definitely in the lake.

"I must have hit that ball pretty hard."

Fred thought *That golf ball must have hit me pretty hard*, but being polite, he didn't say anything about his pain.

Instead, he offered to go into the water and retrieve the ball.

"Oh, I couldn't ask you to do that!" Kitty exclaimed.

Before she could say another word, Fred took off his shirt and headed into the water. He couldn't see the ball in the shallow parts of the lake. He put on his swim mask that he always carried in the back pocket of his jogging shorts. Fred liked to be prepared.**

He took a big breath and swam down into the deeper part of the lake. He could see little fish swimming around. He was hoping there wasn't a great white shark in the Great Lake—or the big fish that swallowed Jonah.

When he was ten feet underwater, he could feel some pressure on his eardrums. Down in the mud below was a rusty anchor, a bunch of clams, and an old desk lamp. There was no sign of Kitty's golf ball.

Fred swam to the surface to get a breath of air. He took off his mask. It had been leaking a little.

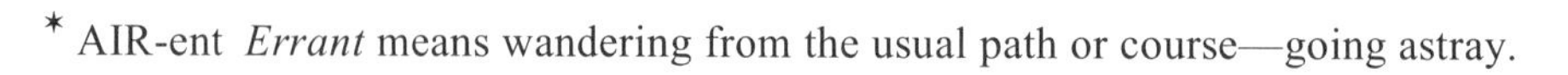
* AIR-ent *Errant* means wandering from the usual path or course—going astray.

** It's good to be prepared. The next time that Fred went jogging, he might wear a helmet to prevent injury from errant golf balls. He would need a helmet with a big space for his nose.

He read the instructions on the side of the mask. That's always a good idea *before* you use something.

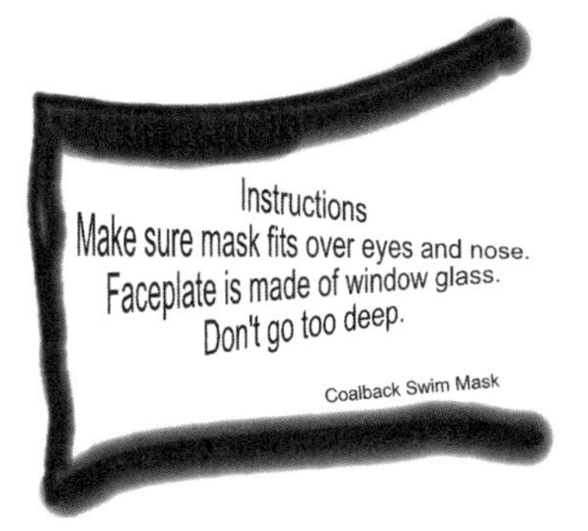

Fred thought to himself No wonder it leaked. I couldn't fit it over my nose.

When Fred put it on again, he adjusted it up really tight.

Really tight. It was so tight that his eyes got pulled around.

He loosened it a bit and things felt better.

Your Turn to Play

1. The swim mask instructions included the words: Don't go too deep. Normally, swim masks don't use window glass. It breaks too easily. If you dive too deep, the pressure on the mask might be too great.

Pressure can be measured in pounds per square inch or pounds per square foot or pounds per square mile. In any event, pressure is defined as force divided by area. (Force and weight measure the same thing.)

Official definition: $\text{Pressure} = \frac{\text{Force}}{\text{Area}}$ $\qquad P = \frac{F}{A}$

How much pressure do you feel if I pat you on the back with my hand (area = 12 square inches) with a force of 5 pounds? (Your answer will be in pounds per square inch.)

2. How much pressure do you feel if I'm using 5 pounds of force, but am accidentally holding a corkscrew in my hand? The point of a corkscrew has an area of 0.01 square inches.

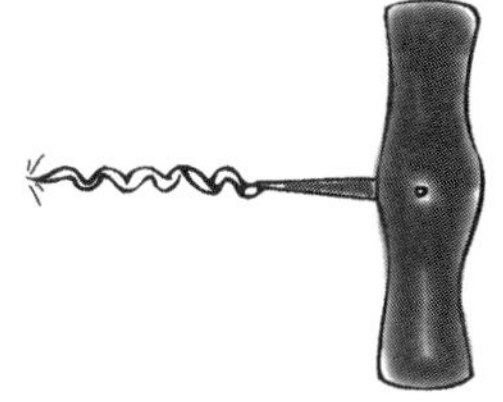

3. Explain why pressure could not be expressed in kilograms per square meter.

.......COMPLETE SOLUTIONS.......

1. Pressure = $\dfrac{\text{Force}}{\text{Area}} = \dfrac{\text{5 pounds}}{\text{12 square inches}}$

$\frac{5}{12}$ pounds per square inch is sometimes written as $\frac{5}{12}$ psi. If you buy a car, it usually comes with tires. On the side of the tires it may read:

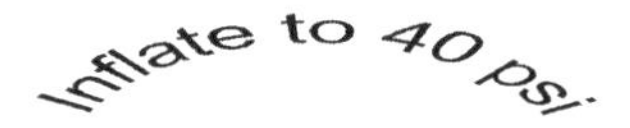

People who work in physics or in tire stores* see *psi* and think *pounds per square inch*. On the other hand, mathematicians see *psi* and think of the penultimate** letter of the Greek alphabet: ψ (pronounced SIGH).

Pounds per square inch can be abbreviated as $\dfrac{\text{lbs.}}{\text{sq. in.}}$ or $\dfrac{\text{lb}}{\text{in}^2}$

There are no periods for abbreviations in the metric system. In the British system things are not quite settled. Sometimes you will see periods and sometimes not, depending on which book you are reading.

2. Pressure = $\dfrac{\text{Force}}{\text{Area}} = \dfrac{\text{5 pounds}}{\text{0.01 square inches}}$ = 500 psi. Ouch!

3. Kilograms are a measure of how much mass—how much matter—something has. It is not a measure of force (or weight).

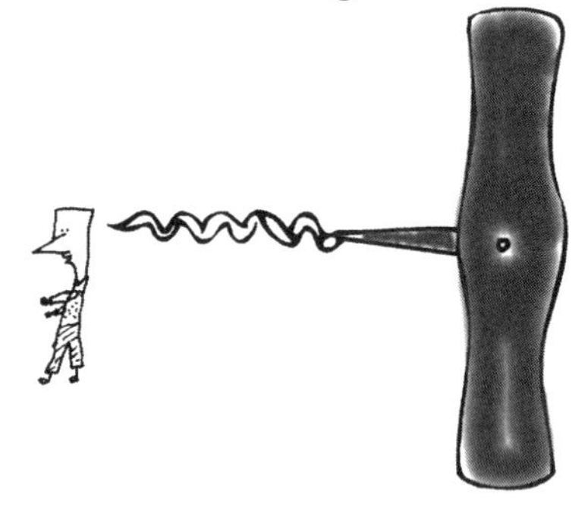

If you were in outer space beyond the Earth's gravity, a 400-kg corkscrew would exert no pressure on you.

However, it would have a lot of inertia. You wouldn't want to walk into one! That would hurt.

---

* Are there physicists who work in tire stores?

** *Penultimate* means next-to-last. There are 24 letters in the Greek alphabet. ψ is the 23rd.

Chapter Twenty-three
Swim Masks

With his mask adjusted properly, Fred dove down into the Great Lake. When he was 15 feet under the surface, he could hear some creaking and crackling in his swim mask. At 17 feet he decided to remove the mask before it imploded* into his face. He found he could swim just as well without the mask.

THE BIG QUESTION

HOW MUCH PRESSURE IS ON HIS MASK WHEN IT IS 17 FEET UNDERWATER?

First thought: Does it matter which direction the mask is facing: left, right, up, or down?

Look at a point 17 feet under the surface of the water. Is the pressure at that point the same in all directions?

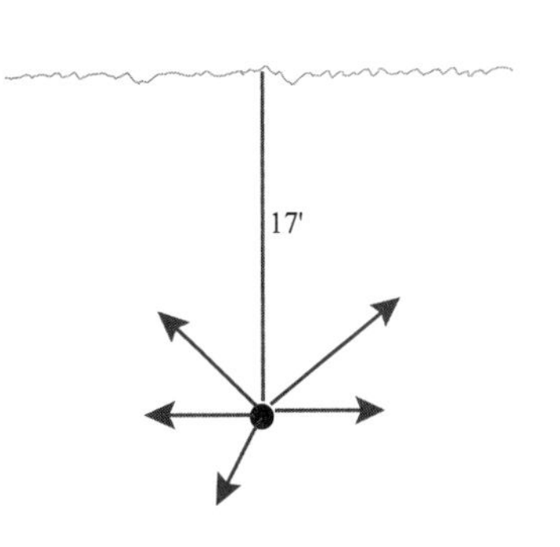

Physicists like to do experiments. If they took Fred and a whole bunch of identical Coalback swim masks, they could plunge him into the water facing left,

facing right,

facing up,

facing down,

---

* *Imploded* means to burst inward. (im-PLO-ded where PLO rhymes with FLOW) It is the opposite of exploded.

and measure the depth at which the mask broke. In each case, they would find that it broke at 17 feet under the surface.

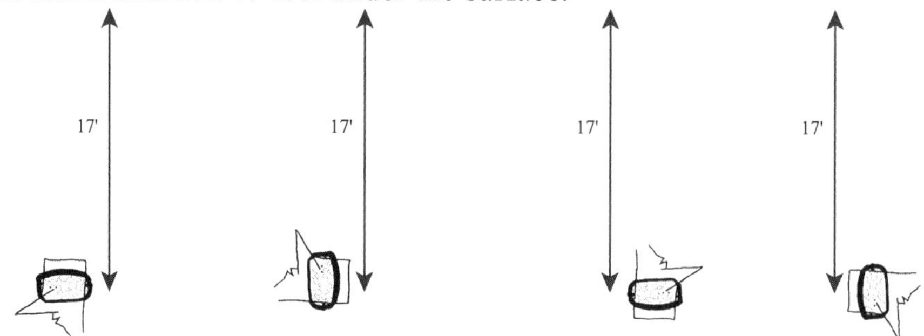

But the SPCF* might object to such an experiment.

If the physicists did the experiment with rabbits, then the SPCR might object.

If they put a swim mask on an ice cream cone, then the SPCICC (of which, I am a member) would object.

Fred wondered *If any point in a fluid didn't have equal pressure in all directions, wouldn't that mean that a cup of coffee would stir itself?*

In any event, physicists have done the experiments (maybe using swim masks on cans of yam soup) and declare: At any point in a fluid, the pressure is the same in all directions.

The BIG QUESTION remains: How much pressure is on his mask when it is 17 feet underwater?

By definition, Pressure $= \dfrac{\text{Force}}{\text{Area}}$

---

* **S**ociety for the **P**revention of **C**ruelty to **F**red

The area of glass in the mask is, say, 10 square inches, so all we need to find is the force. The force is the same as the weight of the water on the mask.

small essay

Areas of Swim Masks

Fred's swim mask is in the shape of an oval, also known as an ellipse. Finding the area of an ellipse is not as easy as finding the area of a rectangle.

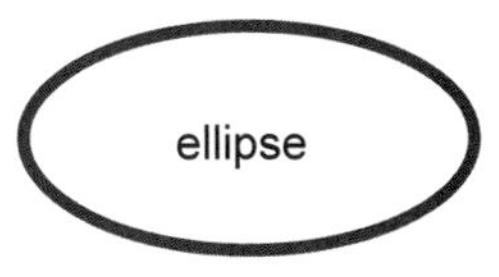

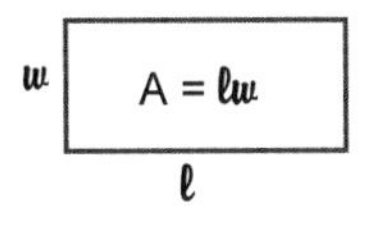

After the high school courses of beginning algebra, advanced algebra, geometry, and trig, you will study college calculus.

In *Life of Fred: Calculus*, on pages 135–136, you will learn both what the area formula for an ellipse and why the formula is true.

Your Turn to Play

1. If the area of his swim mask was 10 square inches and the weight of the water (directly above the mask) was 80 pounds, what would be the pressure on his mask?
2. Does 12 square inches equal one square foot? (Think about drawing twelve squares on a sheet of paper where each square is one inch on every side. Think about drawing one square on a sheet of paper where each side is one foot.)
3. Matching Question. Match each item in the left column with one item in the right column.

pressure	square feet
mass	pounds
area	kilograms
force	pounds per square foot

4. If square inches are sometimes abbreviated as in^2, guess how square feet might similarly be abbreviated.
5. The volume of a golf ball might be 1.7 cubic inches, which can be abbreviated as 1.7 in^3.

Guess how cubic feet might similarly be abbreviated.

.......COMPLETE SOLUTIONS.......

1. Pressure $= \dfrac{\text{Force}}{\text{Area}} = \dfrac{80 \text{ lbs.}}{10 \text{ in}^2} = \dfrac{8 \text{ lbs.}}{\text{in}^2}$ or 8 psi

2. One square foot contains 144 square inches.
Pretend for a moment that this is a square foot:

1 foot

1 foot

If it is divided into square inches,
it would look like this:

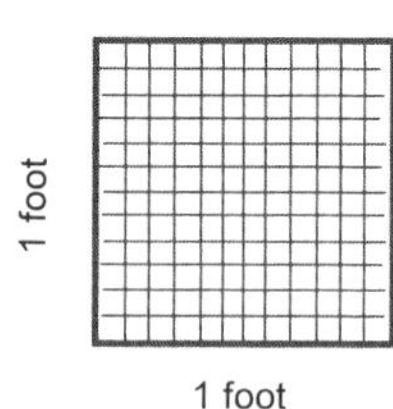

3. pressure matches with pounds per square foot
mass matches with kilograms
area matches with square feet
force matches with pounds

4. Square feet can be abbreviated as ft^2.
5. Cubic feet can be abbreviated as ft^3.

Later on when we get to algebra, we will introduce exponents. Exponents are the little tiny raised numbers.

7^3 is seven with an exponent of three.
7^3 means $7 \times 7 \times 7$.
7^3 is read "seven cubed."

10^2 is ten with an exponent of two.
10^2 means 10×10.
10^2 is read "ten squared."

Exponents come in really handy with big numbers.

10^2 is one hundred	100
10^3 is one thousand	1,000
10^6 is one million	1,000,000
10^9 is one billion	1,000,000,000

Chapter Twenty-four
Density

If Fred were swimming under seventeen feet of cotton candy, there would be a lot less pressure than swimming under seventeen feet of water. Water is a lot denser than cotton candy.

It would be silly to say that water weighs less than cotton candy. It doesn't. A ton of water weighs exactly the same as a ton of cotton candy, but a ton of cotton candy takes up a lot more volume than a ton of water.

The weight density* of water is 62.43 lbs./ft^3.

The weight density of cotton candy is roughly 0.05 lbs./ft^3.

small essay

Least Dense and Really Dense

The least dense place in the universe is in outer space where there is a pure vacuum.** The (weight) density of a vacuum is 0 lbs/ft^3.

One star out of every 10^8 stars in our galaxy is a neutron star. ($10^8 = 10 \times 10 \times 10 \times 10 \times 10 \times 10 \times 10 \times 10 = 100{,}000{,}000$, which is one hundred million) Neutron stars are very dense. A piece of a neutron star the size of a sugar cube would weigh about as much as all humanity.

You couldn't lift a piece of a neutron star that was the size of a grain of sand.

end of small essay

---

* Officially, *density* means mass density (such as kilograms per cubic meter). When writing about pounds per cubic foot, we are dealing with *weight density*. When you see the word *density* in this book, think *weight density*. That is what I should have written.

** Actually, in outer space there is an occasional atom in a cubic foot, but individual atoms are pretty light. For example 6×10^{23} atoms of iron has a mass of 55.8 grams. So one atom of iron would have a mass of

$$\frac{55.8}{600{,}000{,}000{,}000{,}000{,}000{,}000{,}000} \text{ grams,}$$

which is about 0.0000000000000000000000093 grams per iron atom.

Weight density is defined as weight divided by volume. If you were at a fair and bought the really big cotton candy, it would measure 4 feet by 5 feet by 6 feet, and it would weigh 6 pounds.

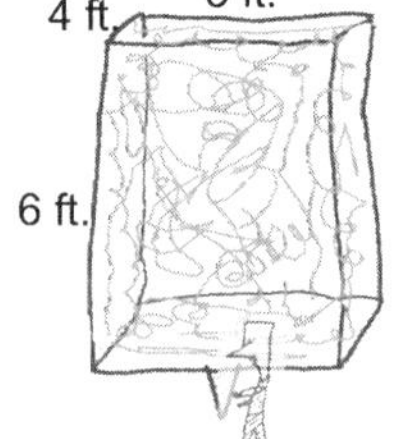

The volume of a box is length times width times height = *lwh* = (4)(5)(6) = 120 cubic feet.

The weight density = $\frac{\text{weight}}{\text{volume}} = \frac{6 \text{ lbs.}}{120 \text{ ft}^3} = 0.05 \text{ lbs./ft}^3$.

Official formula for weight density: $d = \frac{w}{v}$

where w is the weight and v is the volume.

Map of Where We Are and Where We Are Going

The BIG QUESTION is: How much pressure is on his mask when it is 17 feet underwater?

Pressure = $\frac{\text{weight}}{\text{area}}$

and we know the area of the mask is 10 in^2.

Weight density = $\frac{\text{weight}}{\text{volume}}$

and we know that the density of water is 62.43 lbs./ft^3.

What we do not know is the volume of water. That piece of information will unlock everything else.

We want the volume of water over some given area.

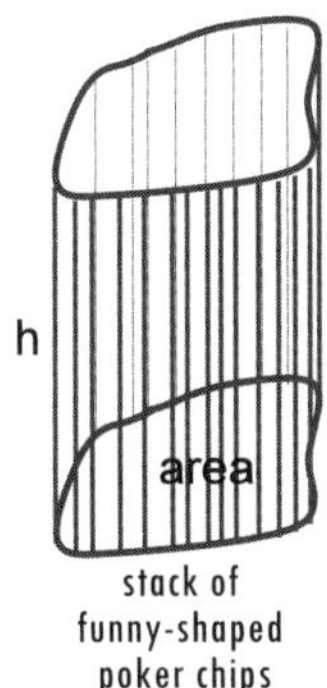

stack of funny-shaped poker chips

The volume of a stack of poker chips is easy to find. It is the area of a poker chip times the height of the stack.* The shape of the poker chip does not matter. If it is in the shape of a circle, you have a cylinder (a tin can). If the poker chip is square, you have a parallelepiped (a cube of butter).

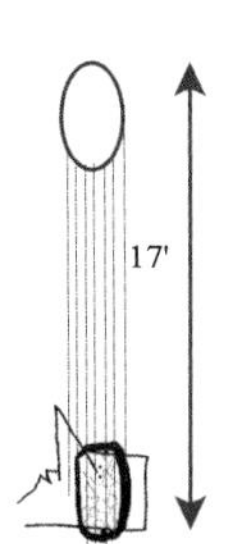

In any event, the volume of water over Fred's swim mask is its area times 17 feet.

We are assuming that the glass of the swim mask is horizontal so that every point on the glass is 17 feet from the surface of the water.

Your Turn to Play to Answer the BIG QUESTION

How much pressure on his mask when it is 17 feet underwater?

We know:

Pressure = $\dfrac{\text{weight}}{\text{area}}$ Weight density = $\dfrac{\text{weight}}{\text{volume}}$ Volume = (area)(height)

$p = \dfrac{w}{a}$ $d = \dfrac{w}{v}$ $v = ah$

1. If $d = \dfrac{w}{v}$ what does w equal?

2. Start with $p = \dfrac{w}{a}$ and move to $p = \dfrac{w}{a} = \dfrac{dv}{a}$ Why is that true?

3. $p = \dfrac{w}{a} = \dfrac{dv}{a} = \dfrac{dah}{a}$ Why is the last equality true?

4. $p = \dfrac{w}{a} = \dfrac{dv}{a} = \dfrac{dah}{a} = dh$ Why is the last equality true?

---

* This is called Cavalieri's Principle. Volume = area of one of the poker chips times the height. V = (area of the base)(height). Cavalieri's whole name was Bonaventura Cavalieri.

No one, except, perhaps, his mother, calls it the Bonaventura Cavalieri's Principle.

I call it the Poker Chip Principle when no one is listening.

.......COMPLETE SOLUTIONS.......

1. Start with $d = \frac{w}{v}$

Multiply both sides by v $dv = \frac{wv}{v}$

Simplify $dv = w$ (You learned about canceling in *Life of Fred: Fractions*.)

2. The question asks why does $\frac{w}{a} = \frac{dv}{a}$?

The answer to question 1, (dv = w), is the reason that $\frac{w}{a} = \frac{dv}{a}$

3. The question asks why does $\frac{dv}{a} = \frac{dah}{a}$?

We know v = ah. Volume = (area)(height). That was the Poker Chip Principle (Cavalieri's Principle).

4. $\frac{dah}{a} = dh$ We canceled the a's.

Question 4 says $p = \frac{w}{a} = \frac{dv}{a} = \frac{dah}{a} = dh$.

Getting rid of all the stuff in the middle, we have p = dh.
Pressure is equal to density times height.

In the *Your Turn to Play*, we started with $p = \frac{w}{a}$ $d = \frac{w}{v}$ and $v = ah$

and arrived at $p = dh$.

This answers the BIG QUESTION.

The pressure on Fred's swim mask is $(62.43\ lbs/ft^3)(17\ ft)$ which is roughly equal to $1054\ lbs/ft^2$.

$$\begin{array}{r} 62 \\ \times\ 17 \\ \hline 434 \\ 62 \\ \hline 1054 \end{array}$$

That is more than a half of a ton!
(2,000 lbs. = one ton)

The Bridge

from Chapters 1–24

first try

Goal: Get 9 or more right and you cross the bridge.

1. Joe had seen car owners wash and wax their cars so he thought he should do that with his boat. He pulled it out of the water and onto the beach.* The hose was connected to a water tower. The top of the water in the tower was 50 feet above the hose. What was the water pressure at the end of the hose? (Let the weight density of water equal 62 lbs./ft^3 to keep the arithmetic simple.)

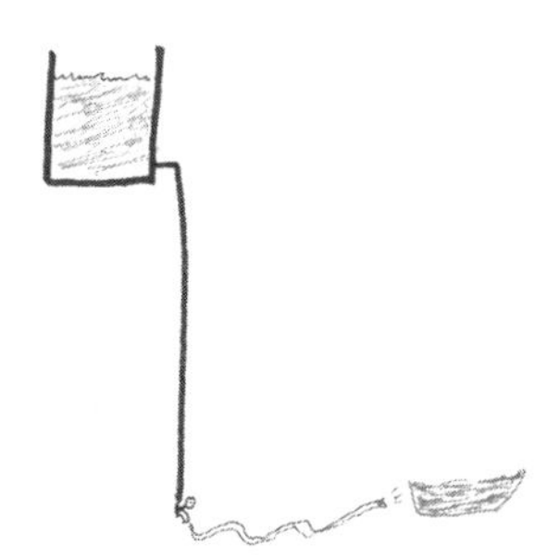

2. One yard equals three feet. How many square feet are in a square yard?

3. $4^3 = ?$ (Work your answer out completely.)

4. Joe sat next to the boat and put a can of wax on his thigh. Then he slowly stood up. Here was the picture when the can first started to slide.
Find the coefficient of static friction between the can and Joe's thigh.

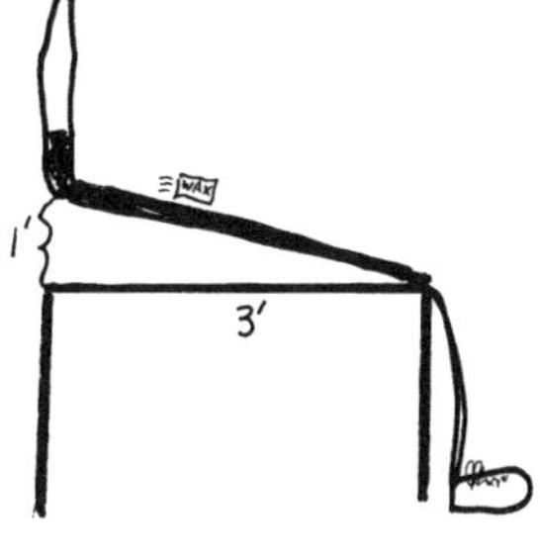

5. Given the information in the previous problem, is it possible to compute the coefficient of sliding friction?

6. $\sqrt{64} = ?$

7. Joe can wax 24 square inches of his boat in 9 minutes. Using a conversion factor, how long would it take him to wax 16 square inches?

---

* Darlene thought that washing a boat was crazy. When the boat was in the water, it was automatically washed. She didn't say anything to Joe. She knew that if she criticized him too much, he would never consent to marry her.

The Bridge

from Chapters 1–24

8. As Joe opened the can of wax, he was converting chemical energy (the energy from the candy and Sluice) into kinetic (motion) energy.

As he rubbed the wax onto his boat, the friction between the waxy rag and the boat was converting kinetic energy into what other form of energy?

9. Joe's can of wax states that it contains 6 kilograms of wax. On the moon where gravity is one-sixth of the gravity on earth, how many kilograms of wax would be in the can?

10. Joe kept track of his success in fishing. One time he spent 1 hour fishing and caught 2 fish. Another time he spent 3 hours and caught 6 fish. Another time he spent 5 hours and caught 10 fish.

He plotted this graph.

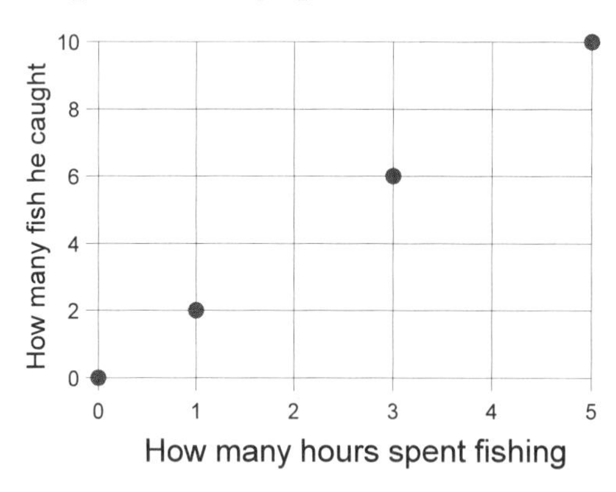

How long would it take for him to catch 4 fish?

The Bridge

from Chapters 1–24

second try

---

1. Darlene was planning her wedding reception. She and Joe were not yet engaged, but she wanted to be ready. There were so many things to consider. She took her *Wedding Planning for the Very Rich* book and opened it at random. She had turned to the chapter entitled, "What Your Guests Will Drink."

It listed a whole bunch of disgustingly expensive wines, scotches, bourbons, beers, etc. Her father once told her, "Why drink something that makes you stupid?"

She decided on bubbly mineral water. It was sold by the cubic foot. She estimated each guest would drink a pint. How many pints in a cubic foot? (1 $ft^3 \approx 7.5$ gallons and 1 gallon = 8 pints.) Use conversion factors.

2. On another page in *Wedding Planning for the Very Rich*, was a description of wedding fountains. The book asked, "How can you have a wedding without a wedding fountain? Won't the guests need a place to throw their coins to show that they are rich?" That all made perfect sense to Darlene. She could gather the coins later to help pay for the wedding.

How much pressure is on a coin that is two feet underwater? (The weight density of water is approximately 62 lbs./ft^3.)

3. At another place in the book she read: "Of course, no wedding reception is complete without a swimming pool for your guests to frolic in." Darlene looked at the pictures of the pools. The smallest pool held 9,000 gallons of water. She figured that she could put Joe in charge of filling up the pool using a garden hose. If the hose could deliver 10 gallons per minute, how many minutes would it take Joe to fill the pool?

4. From the book: "No swimming pool for a wedding reception is complete without adding some goldfish so that the guests will have something to see when they swim underwater." Darlene headed to six pet

The Bridge

from Chapters 1–24

stores and saw thousands of goldfish. All of them were bright orange. She concluded that every goldfish is bright orange. Was this an example of inductive or deductive reasoning?

5. *Wedding Planning for the Very Rich* stated that a large dance floor was an absolute necessity. What would be the area of one that measured 20 meters by 30 meters?

6. If the coefficient of sliding friction between Joe's tennis shoes and the dance floor was 0.7, how much force would it take Darlene to shove him out onto the middle of the dance floor? (Joe weighs 170 pounds.)

7. Darlene would be wearing high heels. She imagined that when she was dancing with Joe, he would twirl her around while she was balancing on one heel. Her square heel measures one-half inch by one-half inch. She weighs 120 pounds. What would be the pressure on the dance floor?

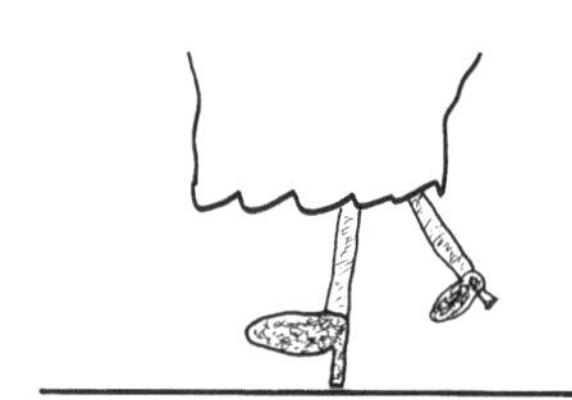

8. If 12y = 7, what does y equal?

9. Is the amount of water in the swimming pool a discrete or continuous variable?

10. *Wedding Planning for the Very Rich* points out that if you are truly wealthy, you will hold the wedding reception in outer space where there is no gravity to bother your guests.

On earth, you could determine the mass of a wedding present (such as a large brick of gold) by weighing it. How do you measure the mass of that large brick of gold in outer space?

The Bridge
from Chapters 1–24

third try

---

1. Fred read some more of Christina Rossetti's poetry. In "Later Life #21" he read:

A host of things I take on trust.

Fred had been told that the gravity on the moon was one-sixth of the gravity on earth, but he had never personally gone to the moon to check that out. Would the density of water (62.43 lbs./ft^3) change if he were on the moon?

2. Fred wanted to buy another copy of *The Works of Christina Rossetti* so that he could give it to Kingie. He went to his desk drawer and got out all his quarters and stacked them up. The stack was 6 inches high. The face of each quarter was 0.8 square inches. What was the volume of that stack?

3. He took one of the quarters and placed it on a slanting surface. Let W be the weight of the quarter. Copy this diagram and do the artwork of resolving W into a normal force and the frictional force.

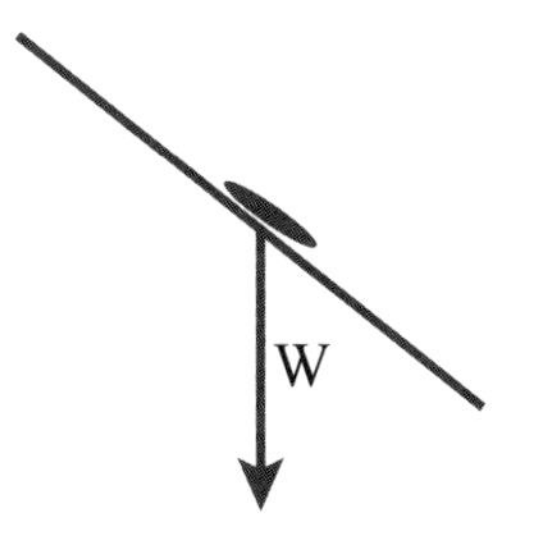

4. Fred's copy of *The Works of Christina Rossetti* weighed two pounds. When he laid it on his desk, how much pressure did the book exert?

5. It took one-half pound of force to push that book across his desktop at a constant speed. What is the coefficient of kinetic friction?

6. How much work is involved in lifting that book three feet off the desktop?

7. How much work is involved in holding that book three feet off the desktop for an hour?

8. Fred read 60 pages of the book in 36 minutes. Using a conversion factor, determine how long it would take him to read 5 pages. (No credit for this problem if you do not use a conversion factor.)

9. $\sqrt{81} = ?$

10. If $0.6 = 0.03x$, what does x equal? (The use of a calculator is not permitted until you get to pre-algebra. It is important at this stage to learn to do the arithmetic.)

The Bridge

from Chapters 1–24

fourth try

1. The three sailors were playing catch with a piece of concrete they had found on one of the streets of New York. When Jane threw it to Jack, would the amount of inertia increase, decrease, or stay the same if she were throwing it to him on the surface of the moon?

2. Would the weight of that piece of concrete increase, decrease, or stay the same if she were throwing it to him on the surface of the moon?

3. Jane lifted the four-pound piece of concrete upward. It went up 5 feet. How much work was involved in that lift?

4. Five feet above her head the concrete stopped moving for an instant. The energy of motion had been converted into what kind of energy when the concrete was held stationary 5 feet in the air?

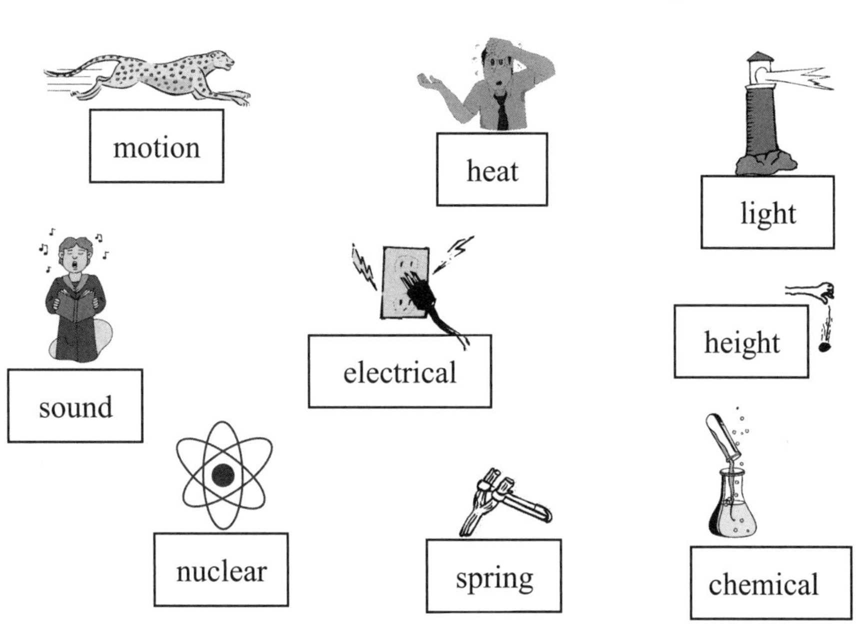

The Bridge

from Chapters 1–24

5. Jake caught the four-pound piece of concrete and placed it on the sidewalk. It took three pounds of force with his foot to get it moving. What is the coefficient of static friction?

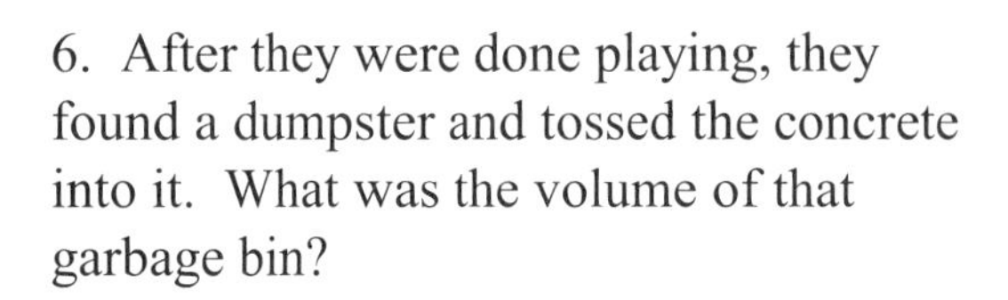

6. After they were done playing, they found a dumpster and tossed the concrete into it. What was the volume of that garbage bin?

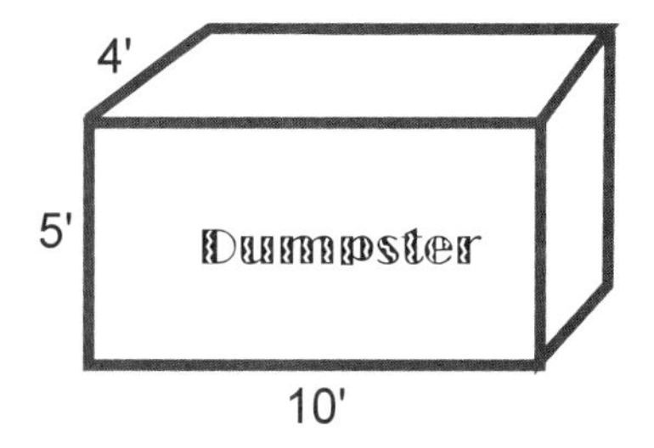

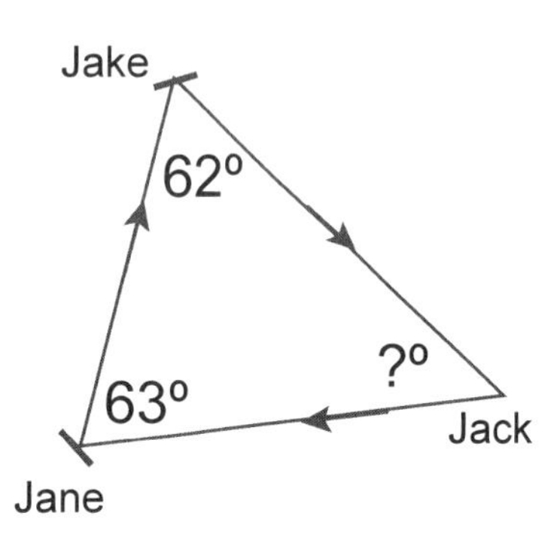

7. Jack likes anything electrical or electronic. He handed Jake and Jane mirrors and they stood in a large triangle. Jack turned on his super-powered flashlight that he had purchased. ("Brighter than six suns!"SM) The light went from his flashlight to Jane's mirror to Jake's mirror and back to his flashlight. How large is the angle at Jack?

8. After a minute, they got tired of playing with the mirrors. Jake put the mirrors and the flashlight into Jack's backpack. He wasn't watching where he was walking and stepped into a bucket of dirty motor oil that someone had left on the sidewalk. The bottom of his shoe was 1.2 feet below the surface. Dirty motor oil has a density of 40 lbs./ft^3. What was the oil pressure on the bottom of his shoe?

9. Jake knew that kicking concrete and submerging his shoe in dirty motor oil would make it hard for his shoes to pass inspection when he got back to the ship. If he earned $8/hour as a sailor, how long would it take him to earn enough to buy a new pair of shoes, which cost $20?

10. If 50x = 25, what does x equal?

The Bridge
from Chapters 1–24

fifth try

Stanthony was inventing a new pizza for his PieOne restaurant: Oily Olive pizza. He took an olive (0.07 lbs.) and lifted it 1.8 feet into the air.

He dropped it into a large bowl of olive oil (density = 36 lbs./ft^3).

In the oil it fell at the rate of 2 inches per second.

1. How much work was done by Stanthony in lifting the olive?

2. How far below the surface would the olive have to fall in order for the pressure on the olive to equal 9 lbs./ft^2?

3. How long would it take the olive to get 7 inches below the surface?

4. He wanted to place an ad for his Oily Olive pizza in the *KITTEN Caboodle* newspaper. The newspaper charges by the square inch. For example, a 6-square-inch ad costs \$8. Using a conversion factor, determine how much Stanthony's 66-square-inch ad will cost. (No credit is given for this problem unless you use a conversion factor.)

PieOne Presents

Stanthony's Newest Creation

Oily Olive pizza

5. Stanthony's Oily Olive pizza comes in various sizes. The most popular is the personal size pizza, which weighs four pounds. How much would that pizza weigh on the moon where the gravity is one-sixth as much as earth?

6. On the first day that he had Oily Olive pizza on the menu, he sold 3 of them. On the second day, he sold 4 of them. On the third day, he sold 8 of them. Graph these three points (where the first coordinate is the number of the day and the second coordinate is the number of Oily Olive pizzas sold.)

7. Is the number of pizzas sold a discrete or a continuous variable?

8. The area of a regular size Oily Olive pizza is 150 square inches. What would be the volume of a stack of these pizzas that was 80 inches high? (Your answer will be in cubic inches.)

from Chapters 1–24

9. A large size Oily Olive pizza weighs 30 pounds and has an area of 6 square feet. What pressure does it exert against the table it is on?

10. $\sqrt{9} = ?$

Chapter Twenty-five
Why Bubbles Float Upward

Fred swam down into a deeper part of the lake. Without the Coalback swim mask he felt a lot safer.* There were lots of things to see. When he looked up, he could see the bottom of boats in the water. He looked at the fish in the water. He could tell that he wasn't a fish. He made bubbles but the fish didn't.

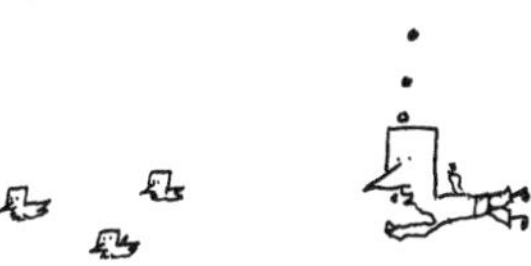

Why do bubbles float upward? Fred thought. *Wouldn't it be nice if they stayed right next to my nose? Then I could breathe them in again and stay underwater a lot** longer.*

In a recent survey, 99.38% of all people have never seriously thought about: Why do bubbles float upward? Why do old desk lamps sink downward? Why do fish neither float nor sink (unless they want to)?

Bubbles have air inside of them. **Most people know that.**

Air is about 80% nitrogen and 20% oxygen. **Some know that.**

Air has mass. **All those who have seen a tornado knock down a house know that air is not nothing.**

---

* Only swim masks made by C.C. Coalback are made from breakable window glass.

In the early days of automobiles, windshields were made of regular glass. This was a mistake. In an accident, those windshields might shatter and—how can I say this delicately? The passengers would be a bloody mess.

Nowadays, all car windshields are made of a special glass that can't shatter. When a modern windshield is broken, the cracks look like a spider web, but there are no splinters of glass flying at you.

This special glass is made by sandwiching clear plastic between two layers of high-strength glass—like two crackers with a ton of peanut butter between them.

** **Alot** is not a word.

Everything on earth that has mass will have a weight.

Balls have weight and fall when you drop them. Eggs have weight and fall when you drop them. Feathers have weight and fall when you drop them. Babies have weight and don't float up in the air when you put them down for a nap.

But bubbles float upward!

Bubbles should fall to the bottom of the lake, but they don't.

What is pushing them upward?

Let's look at a bubble in the shape of a tin can (a cylinder). It's not moving sideways because the force on the left side is equal to the force on the right side.

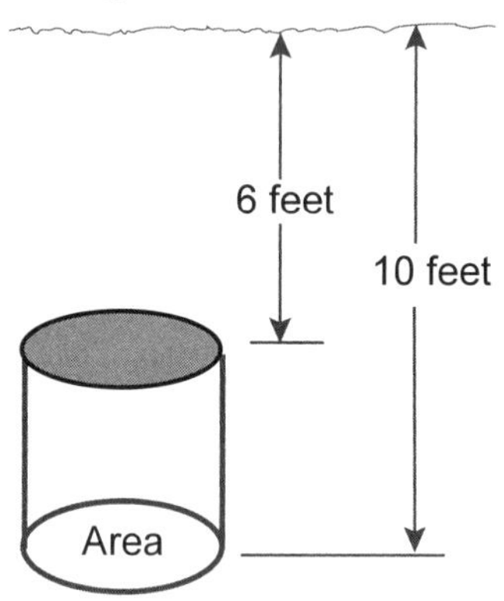

❁ What is the water pressure on the top of the bubble? The formula for the fluid pressure is density times height of the water. $p = dh$

That was the answer to the BIG QUESTION of the previous chapter. So the pressure on the top of the can is 6d, where d is the density of water.

❁ What is the force (or weight) pressing down on the top of the bubble?

Pressure was defined as $\frac{\text{Force}}{\text{Area}}$ or in symbols it can be written as $p = \frac{w}{a}$

Multiply both sides by a and we get $pa = w$. Switch sides and we have $w = pa$. From the previous question, we know that the pressure is 6d. So, finally, $w = 6da$ is the pressure on the top of the bubble.

Wait! Stop! I, your reader, am overwhelmed. I didn't follow all that stuff. You busted my brain! I was happy when Fred was swimming with the fish and blowing bubbles. You didn't lose me when you said that bubbles float upward.

I was even happy with the first ❁, when you found that the pressure on the top of the can was 6d.

But when you got to the second ❁, and did those fancy computations for the weight on the top of the can, my head started to swim.

Let's go v-e-r-y s-l-o-w-l-y over the five lines in the second ❁.

I asked, "What is the force on the top of the bubble?" That question is important because I want to find the force on the top and the bottom of the bubble and compare them.

Then I said that pressure is defined as $\frac{\text{force}}{\text{area}}$ or $p = \frac{w}{a}$ which was given at the bottom of page 126. You worked with that formula in problem 1 on page 128. It was the first formula in the box on page 130. The formula was in problem 2 on page 131. You used the formula in problem 7 of the second try of the Bridges, in problem 4 of the third try, and in problem 9 of the fifth try.

Then $p = \frac{w}{a}$ was turned into pa = w. Here is that computation in slow motion:

Start with	$p = \frac{w}{a}$
Multiply both sides by a	$pa = \frac{wa}{a}$
Simplify	$pa = w$

Then pa = w was turned into w = pa by switching sides. In general, if A = B, then isn't it true that B = A?

We now have w = pa. (Weight is equal to pressure times area.) In the previous ❁, we found that p (pressure) was equal to 6d.

So instead of w = pa, we wrote w = 6da, replacing the p with 6d.

end of long explanation

We will now do the same thing for the pressure on the bottom of the bubble.

❁ What is the water pressure on the bottom of the bubble? The formula for the fluid pressure is $p = dh$. So the pressure on the bottom of the can is 10d, where d is the density of water.

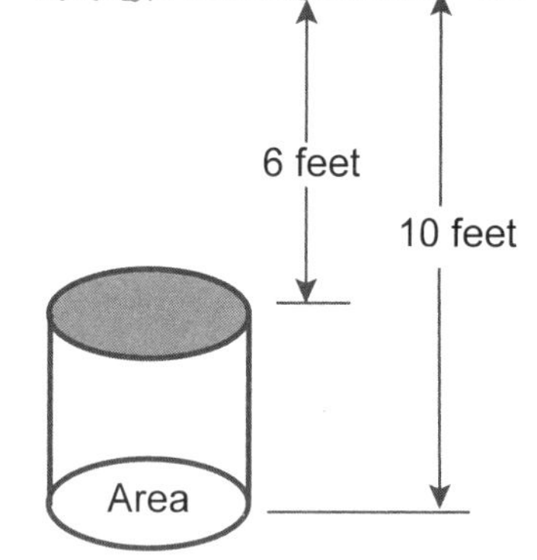

❁ What is the force (or weight) pressing up on the bottom of the bubble?
Pressure was defined $p = \frac{w}{a}$

Multiply both sides by a and we get $pa = w$. Switch sides and we have $w = pa$. From the previous question, we know that the pressure is 10d. So, finally, $w = 10da$ is the pressure on the bottom of the bubble.

Often readers get overwhelmed when they try to read the mathematics at the same speed that they read the adventures of Fred. If they do that, things become **a blur**.

You can't eat steak at the same speed you drink milk.*

In an English class you might be asked to read 30 pages of a novel. In a math class you might be assigned to read 3 pages. Both assignments will take about the same amount of time to complete.

Okay. I'll read the math stuff a lot more slowly than the story parts. Thank you.

❁ ❁ ❁

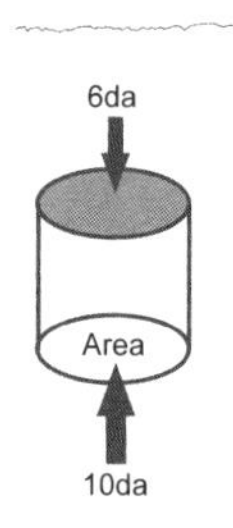

Now, where were we? We found that the force on the top of the can (the bubble of air) was 6da pounds, and the force pushing up on the bottom of the can was 10da pounds. (d is the density of water and a is the area.)

Why is the force on the bottom of the can larger than the force on the top? Because it is farther underwater.

---

* My mother told me that.

If there is a pressure of 6da pounds on the top of the can (the bubble) and a pressure of 10da pounds on the bottom of the can, then the can is being pushed upward by the water with a net force of 4da.

Here's the math:
$10da - 6da = 4da$

That's called the buoyancy* of water = how much the water shoves things toward the surface.

Your Turn to Play

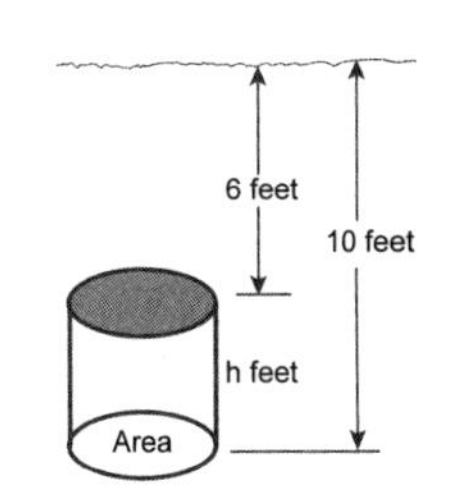

1. In this diagram, how tall is the can?
   In other words, h = ?

2. From the box at the top of this page, we found that 4da is the buoyancy when the can is 4 feet tall.
   In general, if the height of the can is h instead of 4, then buoyancy will equal hda instead of 4da.
   buoyancy = hda for cans that are h feet tall.
   To simplify the equation buoyancy = hda, we ask what does ha equal? (Think Bonaventura Cavalieri from page 131.)

3. Now we have buoyancy of the can = dv.
   d is the density of water. v is the volume of the can.

   What does dv equal when you use weight density = $\frac{\text{weight}}{\text{volume}}$ which is $d = \frac{w}{v}$?

---

* BOY-an-see *Buoyancy*, along with *buoyant* and *buoy* (BOO-ee, the thing that floats in the water as a marker), are three great answers if someone ever asks you, "Can you name a word in the English language that has *uo* in it?" Try that question out on your parents. That is not a *fatuous* (= silly) question.

.......COMPLETE SOLUTIONS.......

1. If one end of the can is 10 feet underwater and the other end is 6 feet underwater, the height of the can will be 4 feet.

2. This is the Poker Chip Principle. The volume of a stack of poker chips, or slices of salami, is equal to the area of any of the slices times the height of the stack. So ha = the volume of the bubble.

3. Starting with the weight density definition ($d = \frac{w}{v}$), and multiplying both sides by v, we get dv = w. The density of an object times its volume is equal to the weight of the object.

The buoyancy of the bubble is equal to dv which is equal to the weight of . . . what?

The volume is the volume of the bubble. But the density is the density of water, not of the bubble.

This is going to be weird.

The buoyancy of the can is equal to the weight of the can *if it were filled with water.* And this force is in the upward direction.

As my father explained it to me: Water does not like to have things shoved into it. Shove a Ping-Pong ball into water and it wants to pop up to the surface. The buoyancy is equal to dv (density of water times the volume of the thing shoved under the water). It just depends on the volume of the thing shoved under the water.

The water says to the thing, "You are in my territory. You are taking up 13 cubic inches of my space and so I'll push you upward with 13d pounds of force.

"If you are 13 cubic inches of Ping-Pong ball or 13 cubic inches of meat loaf, I don't care."

Fancy physics books say, "The buoyancy of an object is equal to the weight of the volume of the water displaced by the object."

Chapter Twenty-six
Why Things Sink

Wait a minute! Before you start Chapter 26, I, your reader, have a big question to ask you.

Cancel this chapter for a moment. We are still in Chapter 25.

Okay.

~~Chapter Twenty-six~~

~~Why Things Sink~~

Now what did you have in mind?

You, Mr. Author, spent six pages talking about why bubbles float upward, but the only bubbles you worked on were in the shape of tin cans. When is the last time you saw a bubble in the shape of a cylinder?

I could draw you some.

Please! Be serious!

Serious? Me? I love writing this book, and I like to be silly, silly, silly, silly, SILLY, silly. When I was a kid in school, I had to sit up straight and not giggle in class. When they gave me tests in school which looked like this:

History 117A

Name_______________

Final Examination

I always wanted to write Name *what?*, but I never did.
I just wrote Name *Stan Schmidt* like I was supposed to.

When I got married in 1992, neither of us especially liked cake. We had pizza instead.

I'm not surprised, given how often you talk about pizza in your books.

But let's get back to those bubbles, which really look more like this:

You draw a lot better than I do.

I'll tell you a secret. Everybody draws better than you do. Wait a minute! You are getting off the topic again. Stay focused. I want to talk about those tin-can bubbles of yours, and you talk about pizza at your wedding.

Okay. I did cylindrical [sill-LYNN-dreh-kel] bubbles since you haven't had beginning algebra,

advanced algebra,

geometry,

trig, and

calculus yet.

It's the only shape I could use to show that buoyancy equals dv (density times volume).

When we get to calculus, we will take your beautiful drawing of a bubble and divide it up into hundreds of tiny cylinders and compute the buoyancy for each of the cylinders and then add them all up.

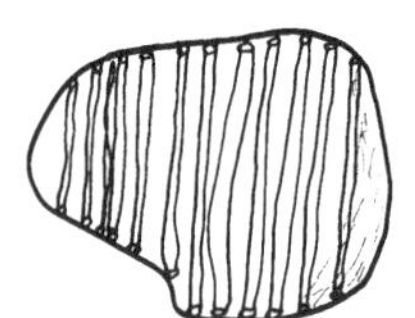

You messed up my drawing! And your lines aren't straight. They are supposed to be tin cans, not spaghetti!

As you said, I can't draw very well. Maybe this is a picture underwater.

My last question: How many tiny cylinders do they use in calculus?

They use so many tin cans that the bottom of each of them has an area smaller than a grain of sand.

That would be more than hundreds of cans.

You are right. In calculus, we use an infinite number of cylinders.

Chapter Twenty-six

Why Things Sink

Fred wondered why the desk lamp at the bottom of the lake didn't float up to the top of the water like his bubbles did. The desk lamp had a volume of 0.08 cubic feet.*

The water buoys up the lamp by dv (density of the water times the volume of the lamp), which is approximately equal to (62 lbs./ft^3)(0.08 ft^3).

The water buoys up the lamp by about 5 pounds.

So, why doesn't the lamp float?
That's easy. It weighs 6 pounds.

$$\begin{array}{r} 62 \\ \times\ 0.08 \\ \hline 496 \end{array} \Rightarrow 4.96$$

Gravity is pulling it downward with a force of 6 pounds. Water is buoying it upward with a force of 5 pounds. Gravity wins.

If someone had thrown a scale into the Great Lake along with the desk lamp, and Fred put the lamp on the scale, it would weigh one pound.

---

* ***I, your reader, have a question. How do you find the volume of a desk lamp? The owner's manual for desk lamps usually does not list that information.***

The first way is to get a barrel and fill it completely with any liquid such as water or Sluice. Then drop the desk lamp into the barrel and measure the amount of liquid that overflows.

But that will make a big mess. And, besides, I would hate to waste the Sluice.

The second way is to get a barrel and poke a hole in the side. Hook a hose up to the hole and run the hose into a big measuring cup. Fill the barrel up to the hole and then plunge the desk lamp into the liquid. The displaced liquid will go through the hose and into the measuring cup.

But Fred wasn't swimming underwater to look for desk lamps. He wanted to find Kitty's golf ball, the one she had bounced off his head and had gone into the Great Lake.

Do golf balls sink or float? Under the official Rules of Golf: a golf ball weighs no more than 1.62 ounces and has a diameter of not less than 1.68 inches.

We need to find the buoyancy of a ball that has a diameter of 1.68 inches and then compare that with a golf ball's weight of 1.62 ounces. If the buoyancy of the ball is less than 1.62 ounces, the ball will sink. If the buoyancy is greater than 1.62 ounces, it will float.

The buoyancy is equal to dv (the density of water times the volume of water displaced by the object). Buoyancy = dv

We know the density of water. $d \doteq 62$ lbs./ft^3 ($\doteq$ means "equal to after rounding off." It is more accurate to say that the density of water is equal to 62.43 lbs./ft^3, but we probably won't need that much accuracy to find out whether golf balls float.)

Using conversion factors, $d \doteq 0.57$ oz./in^3. We needed to change from lbs./ft^3 to oz./in^3 since the Rules of Golf gave the specifications for a golf ball in ounces and inches. All of the yucky arithmetic for the conversion is in the footnote.*

---

* This is a lot of arithmetic. Being a mathematician doesn't mean that I like doing a lot of arithmetic. Some people enjoy arithmetic. I'd rather clean up my office or do the dishes than do a lot of arithmetic. But I'm stuck. I have to figure out whether golf balls sink or float.

First, convert 62 lbs. into ounces. We know 1 lb. = 16 oz.

$$\frac{62 \text{ lbs.}}{1} \times \frac{16 \text{ oz.}}{1 \text{ lb.}} = 992 \text{ oz.}$$

Next, convert cubic feet into cubic inches. 1 ft^3 = 12^3 in^3. Imagine a cube with 12 inches on each edge. Chop it up into little one-inch cubes and there would be 12 × 12 × 12 of those little guys. 12^3 = 1,728

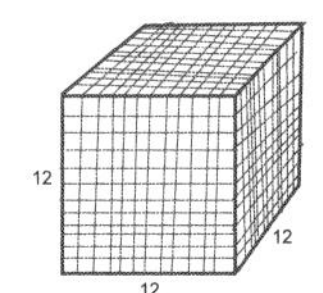

62 lbs./ft^3 = 992 oz/1728 in^3 $\doteq$ 0.57 oz./in^3

We are looking for buoyancy, and we know Buoyancy = dv and d ≐ 0.57 oz./in^3. This is a little like being a detective and following the clues. What we need to know is v, the volume of the golf ball.

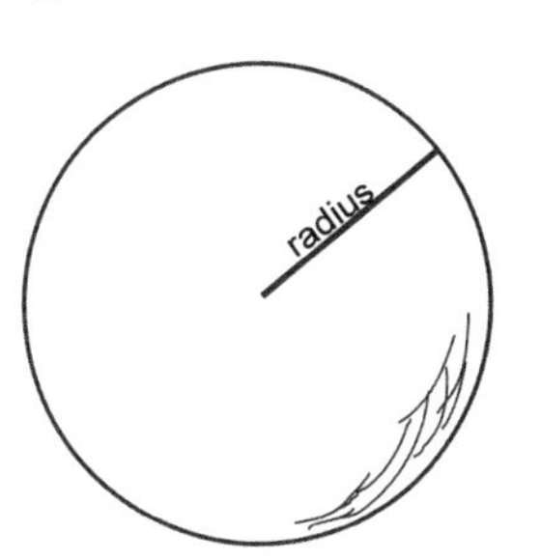

From geometry, we know that the volume of a sphere (a ball) is $v = (4/3)\pi r^3$, where r is the radius of the sphere.
Fact #1: The radius of a ball is one-half of the diameter of the ball.
Fact #2: r^3 means r times r times r.
Fact #3: π is approximately equal to 3.1415926535897932384626433832795.
We are going to round that off a little. $\pi \doteq 3$. ☺

Using Fact #1: radius of a golf ball equals (½)(1.68) = 0.84 inches. (Recall that the Rules of Golf stated that the diameter is 1.68 inches.)

Putting it all together:

Buoyancy = dv = $(0.57)(4/3)\pi r^3 \doteq (0.57)(4/3)3(0.84)^3 \doteq$ 1.35 ounces.

Your Turn to Play

1. What numbers from these last two pages show whether a golf ball will sink or float?
2. A golf ball with a diameter of 5 inches would be a lot easier to hit. The Rules of Golf permit a 5-inch golf ball. Would you want to use one of those big golf balls?
3. You are in the bathtub minding your own business and not bothering anyone. Your younger brother comes in and throws his 80-pound rubber ducky into the water. (That's a heavy ducky.) The volume of that ducky is 2 cubic feet. Will it float?

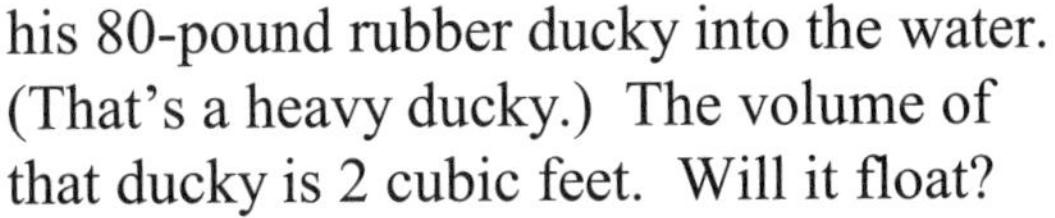

.......COMPLETE SOLUTIONS.......

1. We just computed that the buoyancy of a golf ball is about 1.35 ounces. According the the official Rules of Golf, a golf ball weighs 1.62 ounces. Gravity wins. Golf balls sink to the bottom of the lake.

2. Hee hee. A 5-inch golf ball would be fun to hit but would be impossible to get it into the hole.

3. Buoyancy = dv (where d is the density of water and v is the volume of the water it displaces).

$d \doteq 62$ lbs./ft^3 and $v = 2$ ft^3.

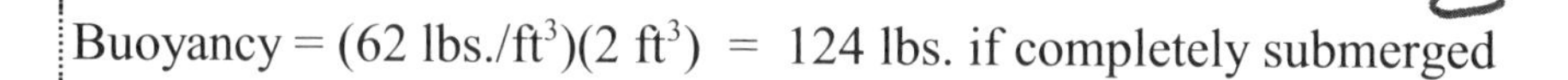

Buoyancy = (62 lbs./ft^3)(2 ft^3) = 124 lbs. if completely submerged

The ducky weighs 80 lbs.

124 lbs. > 80 lbs. It floats. (> means "greater than.")

Now if your younger brother had thrown his 500-lb. steel ducky into your bathtub, things would have been a little different.

If it was still 2 cubic feet, then the buoyancy would still be 124 lbs. since buoyancy just depends on the volume of water displaced.

But now gravity (500 lbs.) would win over buoyancy (124 lbs.) Underwater, that big steel ducky would be pressing on you with a force of 500 – 124 = 376 pounds.

If you let the water drain out of the tub, then the ducky would be pressing on you with its full weight (500 pounds).

Chapter Twenty-seven
Nose and Brain

Fred now knew where Kitty's ball was. It was at the bottom of the Great Lake. He was swimming at the bottom of the lake. There were a zillion golf balls down there. He wasn't sure which one was hers so he gathered them all up, put them into his pants pockets, and swam to the surface.

"Oh, thank Goodness you are okay!" Kitty exclaimed. She had started to get worried about him. He had been underwater for about twenty minutes.*

"There was nothing to worry about," Fred explained. "I could have gone another hour before I needed to come up for air. I have an advantage that most other people don't have.

Official Medical Explanation

Fred has a large air supply that normal people with small noses don't have. His nose can hold gallons of air.

---

* It was exactly 30 pages ago that Fred dove underwater.

Fred can do all kinds of tricks with his nose. If it ever gets in the way, he found that he could just fold it inward. If they had been born with a huge pointy nose, many people would be unhappy. Fred chooses to be happy.

Fred told Kitty, "I wasn't sure which ball was yours so I brought you all of them."

Kitty was delighted. Now she had so many balls that she could practice hitting golf balls all day long before she had to go and pick them up.

"I'm glad to see that the bump on your forehead has gone away," she said. "I was worried that the golf ball may have rattled your brains."

"My brains are well protected," Fred said. "A person's heart and lungs are pretty important, so they are inside the ribs that shield them from injury. You'll notice that you don't have ribs over your tummy." He giggled.

"But my brains are double protected. First of all, they are enclosed inside solid bone. Second, my brains are not packed inside my skull like hamburger is packaged at the grocery store. They are surrounded by brain juice.*

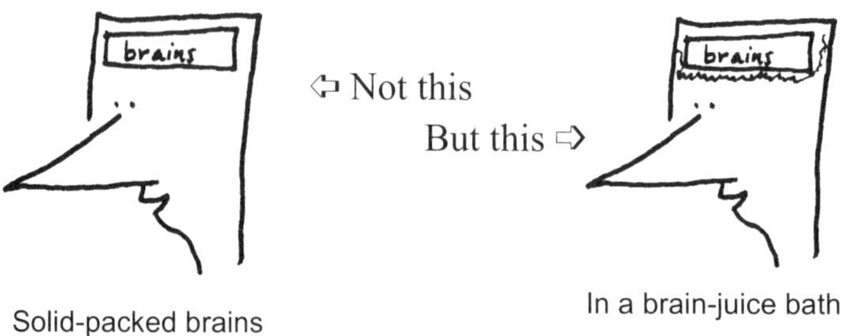

Solid-packed brains

In a brain-juice bath

The density of brain juice is 62.87 lbs./ft^3. The density of brains is 64.93 lbs./ft^3. So the brain is almost weightless in its brain-juice bath. Happy brain.

---

* That's the name that five-year-olds call it. When you get to be six, you call it the cerebrospinal fluid.

Fred put on his socks, shoes, and shirt.

He was about to continue his jogging when Kitty said, "You have been so nice to me. Have you had breakfast yet?"

Fred said, "No." He always told the truth.

Before he could mention that he wasn't hungry since he had eaten a week or two ago, she said, "Fine. I would be happy to buy you breakfast." She pointed to GL&L Café, which was a restaurant located between the Great Lawn and the Great Lake.

Your Turn to Play

1. Let's go back to that 80-pound rubber ducky that your younger brother threw into your bathtub water. The ducky had a volume of 2 cubic feet, and we established in the previous *Your Turn to Play* that it would float.

The ducky does not sit on top of the water. It is not completely under the water. When it is floating, which of these is true?

A) The weight of the ducky is more than the buoyancy of the water.

B) The weight of the ducky equals the buoyancy of the water.

C) The weight of the ducky is less than the buoyancy of the water.

2. What is the volume of the ducky that is under the surface of the water?

(Helpful stuff: Let the density of bath water $\doteq$ 62 lbs./ft^3. The buoyancy formula is Buoyancy = dv, where d is the density of the fluid and v is the volume of the object that is under the water.)

.......COMPLETE SOLUTIONS.......

1. B) If the weight of the ducky were more than the force upward (the buoyancy), then the ducky would sink farther into the water. If the weight of the ducky were less than the buoyancy, the the ducky would be pushed upward.

2. From the previous question, we know that the weight of the ducky (80 pounds) must equal the buoyancy. So 80 = buoyancy.

We know buoyancy = dv, where d is 62 lbs./ft³ and v is the volume of the object that is underwater.

$$80 = \text{buoyancy} = dv = 62v$$

$$80 = 62\,v$$

Divide both sides by 62 $\frac{80}{62} = v$

The rest is arithmetic. $\frac{80}{62} = 1\,\frac{18}{62}$

$$\begin{array}{r} 1\text{ R }18 \\ 62\overline{)\ 80} \\ \underline{62} \\ 18 \end{array} = 1\,\frac{18}{62}$$

$$= 1\,\frac{9}{31}\ \text{ft}^3$$

If we divide top and bottom by 2,
$\frac{18}{62}$ reduces to $\frac{9}{31}$

Please note that part of the goal of this book is to offer one last bit of practice using fractions and decimals without a calculator. The amount of arithmetic in this book has not been extreme—unless you call reducing one fraction by dividing top and bottom by two an extreme amount of arithmetic.

In the next book, *Life of Fred: Pre-algebra 1 with Biology*, you will be able to haul out your calculator if you wish.

Chapter Twenty-eight
The Long Straw

Fred and Kitty sat in the two chairs outside the café. The waiter came and gave them menus. Kitty ordered a hamburger, onion rings, and a strawberry milkshake.

The waiter turned to Fred and waited. And waited. Fred was embarrassed. He really wasn't very hungry. Everything on the menu seemed huge.

Fred asked, "Could I have some water now while I look at the menu?"

After the waiter left, Kitty turned to Fred and said, "After spending twenty minutes underwater in the Great Lake, I'm surprised that you asked for water."

"All the drinks on the menu—milkshakes, Sluice, lemonade—have a lot of sugar in them. I don't want to kill my appetite."

In a minute, the waiter came back with Kitty's hamburger, onion rings, and strawberry milkshake.

He asked Fred, "Do you mind using a straw? We seem to be out of water glasses right now."

Fred didn't understand what the waiter meant, but he answered, "No. I don't mind."*

The waiter handed Fred a big long straw that reached all the way down to the Great Lake.

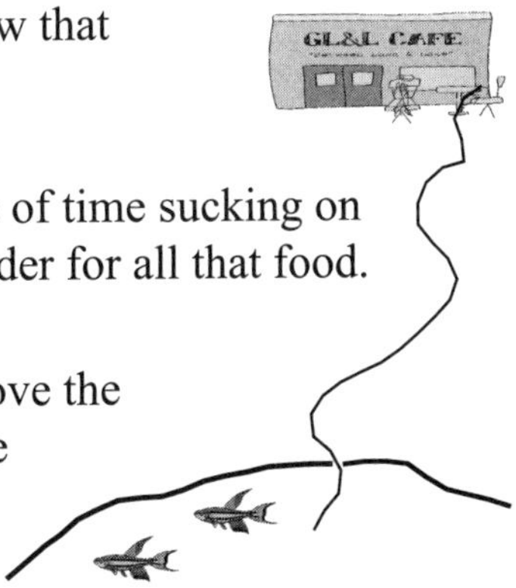

Fred was pleased. He could spend lots of time sucking on the straw, and he wouldn't have to place an order for all that food.

Fred's end of the straw was 35 feet above the surface of the lake. He was about to weigh the atmosphere.

---

* Fred violated a Major Rule of Life: Don't answer questions that you don't understand.

Weigh the atmosphere! I, your reader, object. This is nonsense. Air doesn't weigh anything. Haven't you heard the expression "Light as air"? How can sucking on a straw weigh air?

I was just about to explain before you interrupted me.

Okay. I won't interrupt you.

You just did again!

Sorry.

Apology accepted.

Air does have weight. It has mass. Air isn't nothing. Use a hair dryer and you can feel air.

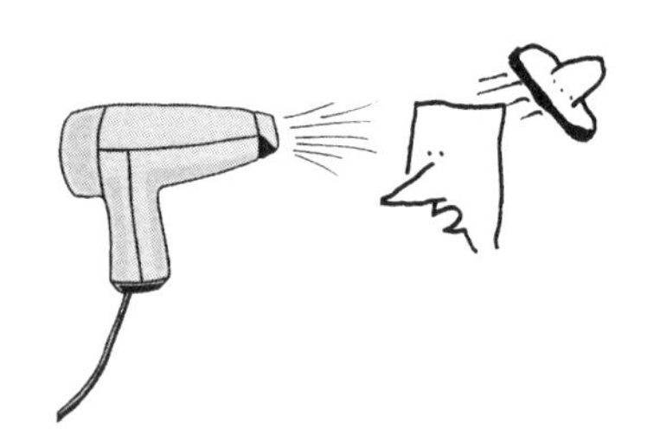

We are swimming in an ocean of air.

Then why can't we float?

We could if we were a lot less dense. The weight density of air is about 0.08 lbs./ft^3. The density of helium gas is about 0.01 lbs./ft^3. That is why helium balloons float upward.

I need to get back to the story. Fred is sitting with Kitty at the GL&L Café sucking on a straw. He is 35 feet above the surface of the lake. The water rises in the straw until it gets up to 34 feet, and then it stops. Fred sucks harder, but nothing happens.

Fact #1: Nothing is caught in the straw.

Fact #2: If Superman were sucking instead of Fred, the water would still only come up to 34 feet above the surface of the lake.

Fact #3: If Fred were to get off his chair and lie on the ground so that his end of the straw was only 33 feet above the surface of the lake, he could drink all the lake water he wanted.

34'

The truth is that sucking on a straw does not make the water, the Sluice, or the milkshake go up the straw. This is something that only people who read physics books know.

When you suck on a straw you make a vacuum. A vacuum is a place where there is no air—no matter of any kind.

And the vacuum sucks up the water.

No it doesn't. A vacuum does not pull the water upward.

I know that you are the author of this physics book, but I don't believe you. I've been sucking on straws for years.

Let's do a thought experiment.*

Suppose we are on the moon where there is no atmosphere and I hand you some Sluice with a straw. Could you drink the Sluice using the straw?

Before you started sucking on the straw, the liquid inside the straw would be at the same level as the liquid in the bottle. Do you agree?

Of course.

When you started sucking, wouldn't you be making a vacuum inside the straw?

That's the whole point of sucking on a straw! You ask silly questions.

But there is already a vacuum in the straw since you are on the moon where there is no atmosphere. Your sucking would not change anything. The liquid would not rise in the straw.

Your Turn to Play

1. Most things expand when you heat them and contract when you cool them. There is one big exception: water. The weight density of water (just before it freezes) is about 62 lbs./ft^3.

After it freezes, its weight density is 92% as much as liquid water.

Compute the weight density of ice. (As usual, please do not use a calculator.)

---

* Albert Einstein was famous for his thought experiments that he used in developing his theory of relativity.

He called them Gedankenexperimente. He spoke German and Gedanke = thought in German.

.......COMPLETE SOLUTION.......

1. We want 92% of 62. (See Chapter 21 of *Life of Fred: Decimals and Percents.*)

of often means multiply.

First change 92% into a decimal. 92% = 92.% = 0.92

Then multiply 0.92 × 62, which equals 57.04 lbs./ft^3 as the density of ice.

$$\begin{array}{r} 62 \\ \underline{0.92} \\ 124 \\ \underline{558} \\ 5704 \end{array} \Rightarrow 57.04$$

Density of liquid water ≐ 62 lbs./ft^3.
Density of ice ≐ 57 lbs./ft^3.

You are alive because this is true.

You are alive because ice is one of the few substances that gets less dense as it gets colder.

You are alive because ice floats.

If ice didn't float, it would sink.

If it sank to the bottom of lakes and oceans, then it wouldn't melt as easily as if it stayed on the surface where the sun can more easily heat it up.

If unmelted ice started accumulating on the bottom of lakes and oceans, it wouldn't be long before we would have to rename the Pacific Ocean as the Pacific Ice Cube.

Now we get to the technical part. There would be no rain. The water in the clouds comes from evaporation from lakes and oceans. Ice doesn't evaporate very well.

No rain → no crops → no cattle → no hamburgers → no you.

(If you are a vegetarian, no rain → no crops → no you.)

Chapter Twenty-nine
Nature Loves Vacuums

Fred is only five years old. He is a professor of mathematics at KITTENS University, but he still thinks that he sucks liquid up a straw. He thinks that the vacuum he creates in the upper end of a straw pulls the liquid up.

How silly!

In the old days, almost everyone thought like Fred. A couple thousand years ago, the common expression was, "Nature hates a vacuum."* They thought that atoms and molecules would somehow sense where a vacuum was and rush to fill it.

If nature hates a vacuum, why is there so much of it? Most of the universe is not stars and planets, but empty space. If vacuums could suck, then our atmosphere would instantly be sucked into outer space.

Instead, the truth is that when Fred sucks on a straw and creates a vacuum, lake water, or Sluice, or milk is *pushed up* the straw.

Cartoon from an early manuscript

---

* Of course, a couple thousand years ago no one spoke English. Even a thousand years ago, there was no one on the planet who could understand you or me or the words in this book.

And the English we speak today will probably not be understandable by people in A.D. 3000. Even your grandparents don't understand a lot of what you say today. And you don't understand them when they talk about shoe hooks and Jeanette MacDonald.

English changes quickly. In contrast, mathematics is much more permanent. It's a pretty good bet that $2 + 2 = 4$ will still be true a thousand years from now.

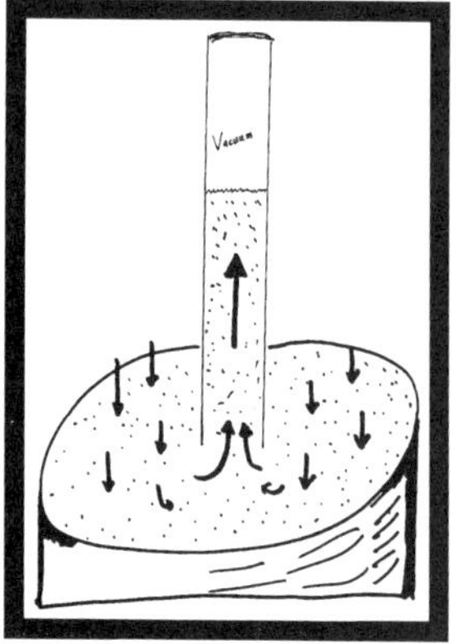

Liquid is pushed up the straw

← Here is the true picture. The atmosphere pushes down on the surface of the liquid and shoves it up the straw.

Experiment Time!

We can learn a lot about nature by looking, by observing, by experimenting. That may seem obvious, but even as recently as Galileo, some people would rather talk about nature than look at it.*

First, we get a glass tube that is 50-feet long and sealed at one end. We fill it with water, invert it, and stick the open end into a lake or dish of water. The water falls until it is 34 feet above the surface of the lake.

We fill a glass tube that is a mile long and sealed at one end. We fill it with water, invert it, and stick the open end into a lake. The water falls until it is 34 feet above the surface of the lake.

Repeat with a 30-foot glass tube. The water doesn't fall at all.

---

* In early Greek times, Aristotle (384–322 B.C.) stated that all the bodies in the sky, including the moon, were perfect spheres. He said that only the earth was imperfect.

Since Aristotle said so, everyone thought he was right.

Then nineteen hundred years pass. In 1609, someone (Galileo) decided to actually lOOk at the moon. What a radical idea!

Did the moon look like a Ping-Pong ball or a billiard ball? No. With his little telescope he saw that the moon had mountains and valleys and plains.

Even if you go out on a clear night and look at the moon without a telescope, you can see that it is not perfectly smooth.

The Big Question is Why did people for nineteen centuries believe the authority (Aristotle) rather than their own eyes and common sense?

Repeat with a big fat tube that is 40-feet tall. The water falls until it is 34 feet above the surface of the lake.

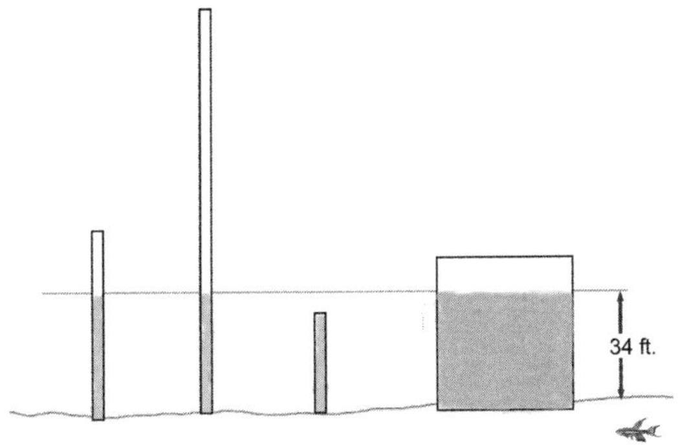

Do you see a pattern? It doesn't matter how much vacuum is in the tube. The air pressure on the surface of the lake pushes the water 34 feet into the tube.

Wait a minute! I, your reader, object. This is like Einstein's Gedankenexperimente. How am I going to get a 50-foot long glass tube?

You could use a 50-foot garden hose.

Um. There's one small problem. How can I tell the water level in the hose? My garden hose is not transparent.

Good point.

Your Turn to Play with a Garden Hose

First, weigh your 50-foot garden hose. Empty hose = 9 lbs.

Then fill the hose completely with water, close both ends, and weigh it again. Hose full of water = 34 lbs.

Then stick one end in a tub of water and open that end.

Hold the other end straight up in the air.

Close the bottom end and weigh the hose. Hose partly filled with water = 26 lbs.

1. Do the math and compute how high the water was in the tube. This will take several steps. For fun, try and do this problem without looking at the hint given below.

If you need help with the math: (1) compute the weight of the water in the full hose; (2) compute the weight of the water in the hose after some of the water has run out of the hose; (3) compute what percent of the hose is filled after some of the water has run out; (4) multiply that percentage times 50 feet.

.......COMPLETE SOLUTION.......

1. If a full hose weighs 34 pounds and an empty hose weighs 9 pounds, then the water in the full hose weighs 25 pounds (34 – 9).

If the hose partially filled with water weighs 26 pounds, then the water that remains in the hose weighs 17 pounds (26 – 9).

$$\frac{\text{weight of water in the partly filled hose}}{\text{weight of water in full hose}} = \frac{17}{25} = 68\%$$

$$\begin{array}{r} 0.68 \\ 25\overline{)17.00} \\ \underline{150} \\ 200 \\ \underline{200} \end{array}$$

$0.68 = 68.\% = 68\%$

Sixty-eight percent of the hose is filled with water.

68% of 50 feet = 0.68 × 50 = 34 feet, which is what we expected.

Nice experiment, but I, your reader, can't figure out how to make a 50-foot hose stand straight up in the air. Any thoughts?

You could take a pole and tie the hose to one end of it and raise it.

I checked with the hardware stores. They don't sell poles that long. Any more thoughts?

After you have filled the hose, find some building that is tall. Carry the hose up the stairs and let one end hang down while holding the other end.

If I'm holding the upper end, how do I put the bottom end of the hose into the water and open up the bottom end?

Do you have any friends?

One end of the hose is easy to close: just screw on a cap or a hose nozzle that is in the off position. How do you close the other end?

A cork.

Thank you.

Chapter Thirty
Weighing the Atmosphere

Kitty had finished her hamburger, onion rings, and a strawberry milkshake. She had offered one of the onion rings to Fred, but he had been busy sucking up the lake water. He had found that the water would only go 34 feet up the straw.

How much pressure does a 34-foot column of water exert? The formula was p = dh (where p is pressure, d is density of the fluid, and h is the height of the fluid). p = dh was the answer to the 

at the end of Chapter 24.

The pressure at the bottom of the straw = (62.43 lbs./ft^3)(34 ft). Fred figured this out in his head. His answer, after rounding off, was 2,123 lbs./ft^2.

What Fred did in his head:

$$\begin{array}{r} 62.43 \\ \times\ \ 34 \\ \hline 24972 \\ 18729\ \ \\ \hline 212262 \end{array} \Rightarrow 2122.62 \doteq 2123.$$

That meant that the air pressure on the surface of the water was equal to the pressure of the column of water.

That meant that the air pressure was 2,123 lbs./ft^2. That is over a ton of pressure on every square foot of anything at the GL&L Café. If you measured the area of Fred's back, you would have to do it in square inches, but there was at least one square foot of area on Kitty's back.

A ton of air pressure on her back? That's crazy. That would crush her and break all of her ribs. Fred must have made an error in his computations.

✓ Is the formula p = dh true? Yes.

✓ Is the density of water roughly equal to 62.43 lbs./ft^3? Yes.

✓ Is the maximum height of water in the straw 34 feet? Yes.

✓ Did Fred make an error in math? Ha. Ha. Don't be silly.

Okay. I give up. Where's the error?

That is simple. It's your error in thinking there is an error.

There are 2,123 pounds of air pressure on every square foot of Kitty. There are 14.74 pounds of air pressure on every square inch of Fred.*

Fred's forehead would be dented in . . . unless there was an equal and opposite pressure pressing out.

In fact, all the sides of his head would be crushed.

But what would happen if you flew Fred into outer space where there is no air pressure?

Then the outward pressure of Fred's body would not be balanced by the atmospheric pressure.

But what would happen?

---

*

Using the conversion factor of $\frac{\text{1 square foot}}{\text{144 square inches}}$ we can convert lbs./ft^2 into lbs./in^2.

$$\frac{\text{2123 lbs.}}{1\ \text{ft}^2} \times \frac{1\ \text{ft}^2}{\text{144 square inches}} \doteq 14.74\ \text{lbs./in}^2$$

Notice that the ft^2 canceled.

```
        14.743
144) 2123.000
     144
      683
      576
      1070
      1008
        620
        576
         440
         432
           8
```

14.743 ≐ 14.74

You guessed it . . .

He would be like a kernel of corn that had popped.

Your Turn to Play

1. Suppose your older sister gave you a balloon. You took the balloon and blew a tiny amount of air into it and tied it.

She might yell at you, "You wasted the balloon!"

You mail the balloon to your friend who lives on the moon—not in a rocket ship, but out there on the surface of the moon where there is no atmosphere. Will she be happy with the balloon?

2. If she lets go of the balloon, will it float upward or fall downward? (The gravity of the moon is about one-sixth that of the earth.)

Happy Fish as drawn by the author

3. A happy little fish is swimming a mile (5,280 feet) under the ocean surface. If ocean water has a density of 62 lbs./ft^3, how much water pressure does the fish experience?

4. Does the pressure on the happy little fish depend on how wide the ocean is?

Happy Fish as drawn by Kingie who is a real artist

5. Kingie's drawing of "Happy Fish" sold for \$600. When the author's "Happy Fish" was offered for sale, no one offered to pay anything for it. Finally, his mother gave him \$12 for it.

Twelve dollars is what percent of \$600?

6. The author wrote a long letter to his mother, thanking her for her purchase of his art. The letter was so long that it required a stamp that was 5% of the \$12 she had spent on his art. How much did that stamp cost?

.......COMPLETE SOLUTIONS.......

1. On the earth, the air pressure on everything, including the outside of a balloon, is about 14.74 lbs./in^2.

On the moon there is no atmospheric pressure.

The balloon will inflate. Your friend will be very happy with her new balloon.

2. Nothing floats upward in an airless environment, not even helium balloons. There is nothing to push it upward.

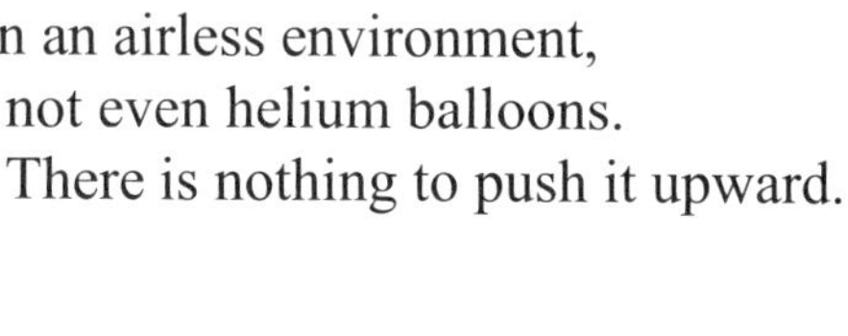

3. p = dh (p is pressure, d is density of the fluid, and h is the height of the fluid). p = 62 × 5280 = 327,360 pounds per square foot!

(There were three aspects to this problem: (1) using the p = dh formula, (2) realizing that there is almost a third of a million pounds of pressure per square foot on creatures that are a mile under the surface of the ocean, and (3) to get a little practice with the multiplication tables. This is the last book before you are allowed to grab your calculator. This is the best time of your life to learn your tables.

$$\begin{array}{r} 5280 \\ \underline{62} \\ 10560 \\ \underline{31680} \\ 327360 \end{array}$$

4. The pressure only depends on the density of the fluid and the depth of the object below the surface. That is what p = dh means.

5. 12 is what percent of 600? When you don't know both sides of the *of*, you divide the number closest to the *of* into the other number. (Chapter 27 of *Life of Fred: Decimals and Percents*)

$$600 \overline{)\,12.00} = 0.02 \qquad \text{with } 1200 \text{ subtracted}$$

0.02 equals 2.% or 2%.

6. 5% of 12. When you know both sides of the *of*, you multiply. (Chapter 21 of *LOF: D&P*) 0.05 × \$12 = \$0.60 or 60¢.

The Bridge

from Chapters 1–30

first try

Goal: Get 9 or more right and you cross the bridge.

1. Joe decided that if he wanted to be a real fisherman, he would have to go where the fish are. He was delighted when he saw an ad in the *KITTEN Caboodle* newspaper: USED SUBMARINE FOR SALE. ORIGINALLY COST $12,000. WILL SELL FOR $360.

The sale price was what percent of the original cost?

2. When Joe went to see the submarine, he noticed that there were two patches on it. He asked the seller, Captain Coalback, about those patches.

Coalback replied, "It's fine. They cover torpedo holes that were made in 1917. [World War I was 1914–1918.] Don't you worry. Those leaks have been almost fixed."

The English was too complicated for Joe.* He smiled and paid for the submarine.

Joe realized that he couldn't get the sub into his car. He tied a rope on the sub and dragged it behind his car. The sub weighed 800 lbs. It took a force of 600 lbs. to get it moving down the street. What is the coefficient of static friction between the sub and the street?

3. (Continuing the previous problem) Suppose Joe threw 200 lbs. of comic books into the sub before he started the car. How much force would it now take to get the sub moving?

4. When he got the sub to the beach, he untied it and rolled it over to inspect the bottom. There was a rectangular hole in the bottom and the edges of the hole felt hot. The rectangular hole measured 0.8 meters by 0.75 meters. What was the area of the hole?

---

* Suppose you are swimming out in the ocean. A giant squid attacks you, and you call for help. The lifeguard swims out to help you. Would you rather be almost drown or nearly saved?

The Bridge

from Chapters 1–30

5. Why did the edges of that hole feel hot? (Please answer in a complete sentence. Don't just write one or two words.)

6. Joe had dragged his sub 3 miles. It took him $\frac{1}{6}$ of an hour. (10 minutes) How fast was he driving? Give your answer in miles per hour.

7. Joe thought it would be fun to put on a rubber suit and go out and catch fish with his hands. He could brag to his friends that he had *really* caught fish.

He knew that he needed to breathe.

He attached a flexible straw to his helmet and the other end to a buoy.

If he were 10 feet underwater what would be the pressure on his chest? (Assume the density of water is 62 lbs./ft^3.)

8. The buoyancy of water will tend to make the submarine float. The weight of the submarine will tend to make it sink.

Is the buoyancy of water on the submarine the same when it is 20 feet underwater as when it is 40 feet underwater?

9. Joe slides the submarine down the ramp and into the water. What is the slope of the ramp?

16 ft.

100 ft.

10. $\sqrt{36} = ?$

---

* If you answered, "Nearly saved" in the footnote on the previous page, that would mean that you are dead.

The Bridge

from Chapters 1–30

second try

1. Darlene was planning her honeymoon.* She and Joe were not yet engaged, but she wanted to be ready.

She thought about going to the country of Freedonia where they use gold coins. Using a conversion factor compute how many gold coins they could get for $111. (Fifteen gold coins are worth $555.)

2. The Freedonia gold coin weighs 0.06 pounds. What is the work done in picking up a gold coin off the sidewalk and putting it in your pocket? (Your pocket is 3 feet above the sidewalk.)

3. How much would a Freedonia gold coin weigh on the moon where the gravity is one-sixth of earth's gravity?

4. The mass of a Freedonia silver coin is 0.18 kilograms. What would be its mass on the moon?

5. On her honeymoon, Darlene wanted to try Freedonia's famous Golden Nectar drink. She wanted to use a really long straw so that everyone could see her enjoy the drink. Freedonia is up in the mountains where the atmospheric pressure is 2000 lbs./ft^2. Golden Nectar has a density of 100 lbs./ft^3. What is the longest straw she could use?

6. Golden Nectar is made from 10 liters of pineapple juice, 4 liters of liquid bee pollen, and 6 liters of boysenberry juice. Boysenberry juice is what percent of Golden Nectar?

7. A Freedonia gold coin weighs 0.06 pounds. It has a volume of 0.0004 cubic feet. How much would it weigh if it were submerged in Golden Nectar that has a density of 100 pounds per cubic foot?

---

* One word in that sentence—**Darlene was planning her honeymoon**—should tell you a lot about Darlene. Which sounds better: *her* wedding, *her* honeymoon, *her* marriage OR *their* wedding, *their* honeymoon, *their* marriage?

8. It takes 0.02 pounds of force to slide a 0.06-pound Freedonia gold coin across a table at a constant speed. What is the coefficient of kinetic friction?

9. The coefficient of static friction for a 4-pound Freedonia copper coin on a table is 0.4. What force is needed to start that coin sliding?

10. If $16x = 2$, what does x equal?

The Bridge

from Chapters 1–30

third try

---

1. Fred was in the bathtub reading some more of Christina Rossetti's poetry. In her poem "Goblin Market" he read:

Like a lily in the flood,
Like a rock of blue-veined stone
*Lashed by tides obstreperously**

The sponge floating in the bathtub water reminded Fred of "a lily in the flood." If the density of water is about 62 pounds per cubic foot, what can you say about the density of the sponge?

2. Fred's big toe on his right foot reminded him of "a rock of blue-veined stone." He pushed his foot under the water. What was the buoyancy on his toe? (Density of bath water = 62 lbs./ft^3. Toe is 6 inches underwater. Volume of toe = 0.0006 ft^3.)

3. He rubbed his big toe against the end of the tub. It made a squeaking sound. He was converting the energy of motion (kinetic energy) into sound energy. The chemical energy in his body was converted into kinetic energy when he moved his foot.

1.2 Calories (of chemical energy) can be converted into 5 hours of toe-squeaking sound. Using a conversion factor, compute how many Calories it would take to produce 3 hours of sound.

4. Fred put his soap on a board and tipped the board until the soap began to slide down into his tub.

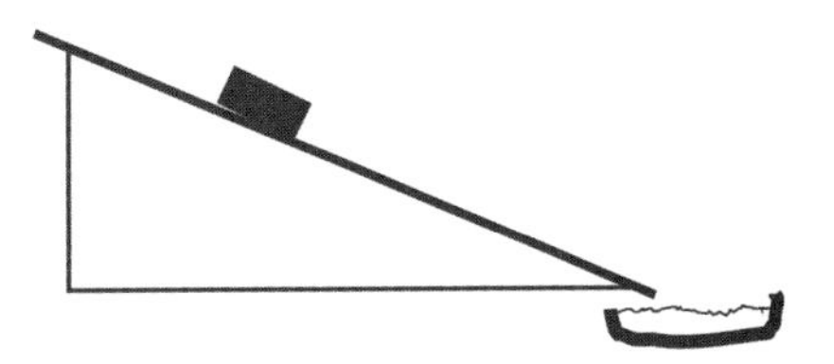

The soap began to slide when the slope of the board was $\frac{3}{4}$. Find μ_s.

---

* ob-STREP-er-ous *Obstreperous* has two meanings: (1) hard to control and (2) noisy.

If you have obstreperous hair, obstreperous would have the first meaning. If you are in an obstreperous factory, you will probably want to wear ear plugs to protect your hearing.

Obstreperous children can be either or both.

The Bridge

from Chapters 1–30

5. Fred's big toe weighs 0.37 pounds. He weighs 37 pounds. What percent of Fred's weight is his big toe? (Hint: The answer is *not* 0.01%.)

6. Fred's bath towel is 27 inches by 60 inches. What is its area?

7. Water was draining out of his tub at the rate of 2 gallons per minute. How long would it be before there was no water in the tub? The tub started with 20.2 gallons in it. Use a conversion factor.

8. If Fred put his big toe 18 inches (= 1.5 feet) under the water, what would be the pressure on his toe? (Density of bath water = 62 lbs./ft^3. Volume of toe = 0.0006 ft^3.)

9. With his wet finger, Fred drew a triangle on the side of the tub.

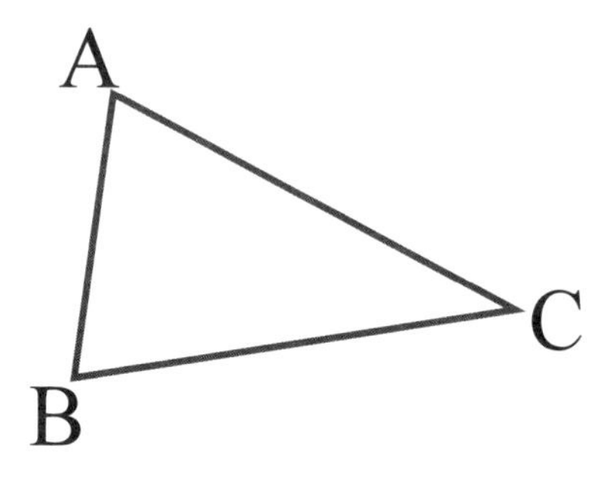

Angle A was 70°.
Angle B was 70°
How large was angle C?

10. $\sqrt{100}$ = ?

The Bridge

from Chapters 1–30

fourth try

---

1. Three sailors were visiting New York. This was their fourth day in the city. Jake said, "It has been sunny every day that we have been here. I bet that it is always sunny here."

Is this an example of inductive or deductive reasoning?

2. Ten minutes later it began to rain. Jane said, "You thought it would always be sunny here. It is raining right now. It is not sunny. Therefore, you were wrong."

Is Jane using inductive or deductive reasoning?

3. The sidewalks were wet and greasy. Using 36 pounds of force, Jane and Jake could push Jack along the street at a constant speed. Jack weighs 180 pounds. What is the coefficient of kinetic friction between Jack and the sidewalk?

4. Jack likes electronic gadgets. He showed Jane and Jake his new Electro-Gizmo™. It was the size of a fat postage stamp: Length = 0.7"; Width = 0.4"; Thickness = 0.1". What was the volume of this little box?

5. His Electro-Gizmo™ could:

➻ Telephone ➻ Tell time ➻ Connect to the Internet ➻ Take pictures ➻ Send faxes ➻ Play music ➻ Vibrate so that Jack could give shoulder massages ➻ Heat up to keep Jack's hands warm in cold weather ➻ Make chocolate milkshakes ➻ Serve as a bookmark ➻ With 51 of its "friends" it could serve as a small deck of cards ➻ Predict the next lunar eclipse ➻ Find the keys that Jack lost ➻ Evaluate a used car that Jack is thinking of purchasing ➻ Warn you if you are in a tornado zone ➻ etc.

Jane asked Jack, "How many of the 40 features of your Electro-Gizmo™ have you used?"

He said, "I've used 15 of them."

What percent of the features had Jack used?

6. Jane said, "I hope your Electro-Gizmo™ is waterproof."

Jack asked, "Why do you say that?"

"It just fell into that giant puddle."

His Electro-Gizmo™ was under 6 inches (= 0.5 feet) of water. The density of puddle water is 62 lbs./ft^3. What was the pressure on his Electro-Gizmo™?

7. Jack reached into the puddle and grabbed his Electro-Gizmo™. He dried it off with his handkerchief. How much work did he do in holding it

steady for five minutes in his hand as he looked at it? His Electro-Gizmo™ weighed 0.08 pounds.

8. In the four days that the sailors had been in New York, Jack bought lots of electronic things.

On the first day, he bought 3 things. On the second day, 2 things. On the third day, 4 things. On the fourth day, 2 things. Graph these four points where the day is the first coordinate and the number of electronic things he bought is the second coordinate.

9. His Electro-Gizmo™ wasn't waterproof. It was completely ruined. Jack held it above the trash bin and released it. He was converting which form of energy into which form?

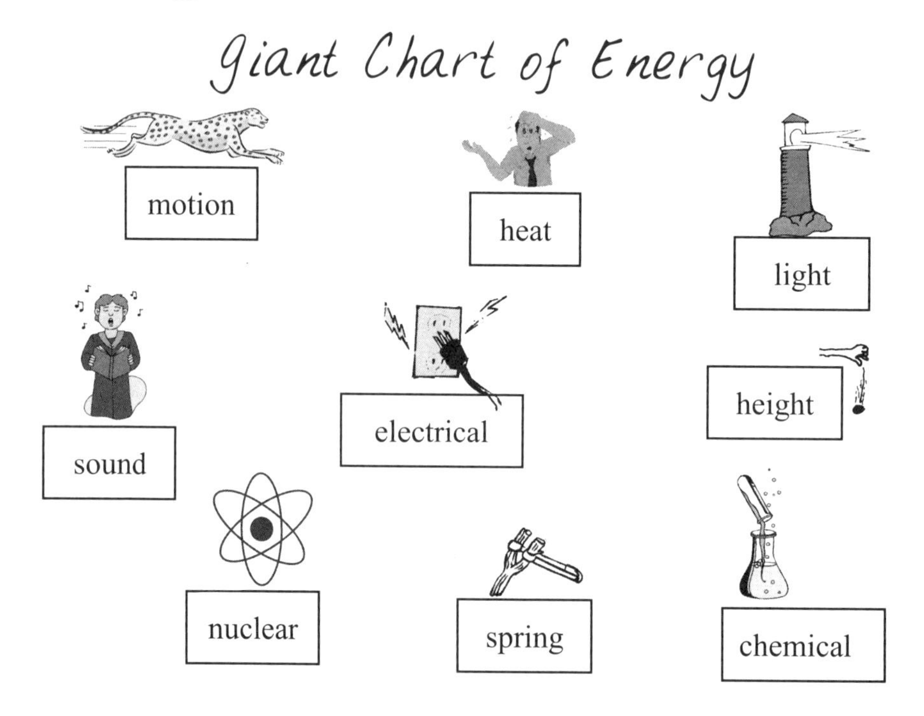

10. $\sqrt{4} = ?$

The Bridge

from Chapters 1–30

fifth try

---

1. Stanthony invented his new **Crispy Pizza**. It was in the shape of a rectangle. The single serving size was 9" by 15". What was the area of that pizza (in square inches)?

2. After cutting the dough, he dropped it into a big tub of hot oil. The density of the oil was 40 lbs./ft^3. The dough fell to the bottom of the tub. What can you say about the density of the dough?

3. What was the pressure on the dough? (It was under 4 feet of oil.)

4. In 4 minutes, the dough was 60% done. Using a conversion factor, compute how long it takes to fully cook the **Crispy Pizza** dough in the oil.

5. After the dough is done, Stanthony uses a large fork to pull it out of the oil. The dough is now like a cracker.

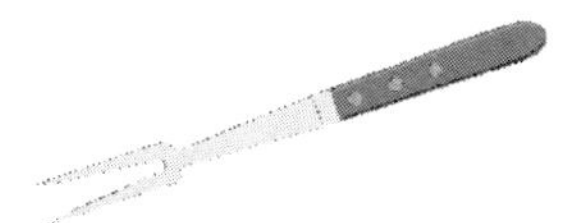

He pushes the crust 3 feet to move it across the table to the place where he can add the toppings. He pushes with a force of 2 pounds. How much work was done?

6. If it takes 2 pounds of force to keep that crust moving at a constant speed, and if the coefficient of kinetic (moving) friction is 0.4, what is the weight of that crust?

7. Stanthony liked to experiment with different toppings for his **Crispy Pizza**.

One week he used pepperoni, green onions, and olives. Eighty-four percent of the customers that he talked with said they liked that.

Another week, he used radishes, pickles, and yams. Seven percent of the customers that he talked with said they liked that.

Yuck!

and 93% said that they hated it

He concluded that the pepperoni, green onions, and olives was the more popular combination. Was this an example of inductive or deductive reasoning?

8. After putting on the toppings he lifted the 6-pound pizza 2 feet upward so that he could slide it into the pizza oven. How much work was involved?

9. Here is a graph that Stanthony made that plotted how many **Crispy Pizzas** were sold each week. How many **Crispy Pizzas** were sold during his best week of sales?

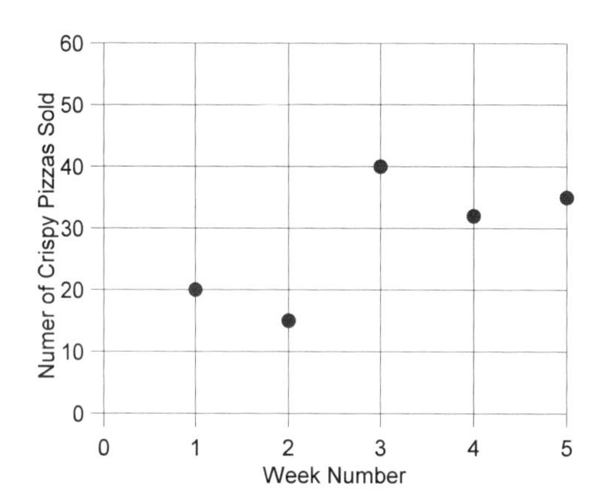

10. If 5y = 30, what does y equal?

Chapter Thirty-one
Water Fountains

The check came. Fred wasn't sure what the correct etiquette* would be in this situation. He offered to pay for half of the bill. Kitty said that she would take care of the check, noting that she had offered to buy him breakfast.

GL&L CAFE

hamburger	3.00
onion rings	2.50
strawberry shake	2.75
water	no charge
Total	8.25

Fred thanked her.

There was a beautiful fountain and waterfall near the GL&L Café. Kitty and Fred stopped to look at it.

Kitty enjoyed the sound of the rushing water. She liked the sight of the white water in the fountain against the background of the green trees.

Fred's eyeballs are five years old. (That's because Fred is five years old.) He could see things that people with older eyeballs couldn't see.

He spotted a tiny dead bug in the water. It floated downstream, dropped over the waterfall, and disappeared into the drain at the bottom of the pond.

A second later, the bug came up the fountain, floated down the stream again, dropped over the waterfall, and disappeared into the drain.

Kitty couldn't figure out what Fred was looking at. She asked, "You really like the fountain, don't you?"

"This is neat," Fred said. "Look! Here comes the bug again."

---

* ETT-eh-kit You can guess by the spelling that *etiquette* was a word we borrowed from the French. Etiquette is the rules of behavior in social situations. For example, at a restaurant you don't wipe your mouth with the tablecloth.

Etiquette depends on which social group you are in. If you are with a group of three-year-olds and you belch so hard that you knock the pictures off the walls, they might think that you are trying to entertain them. No apology is required. It is different in the company of adults.

"What bug?"

Fred reached into the water and got the bug. He put it on the end of his finger and showed it to her. "It's that black dot in the middle of my finger."

"What dot?" she asked.

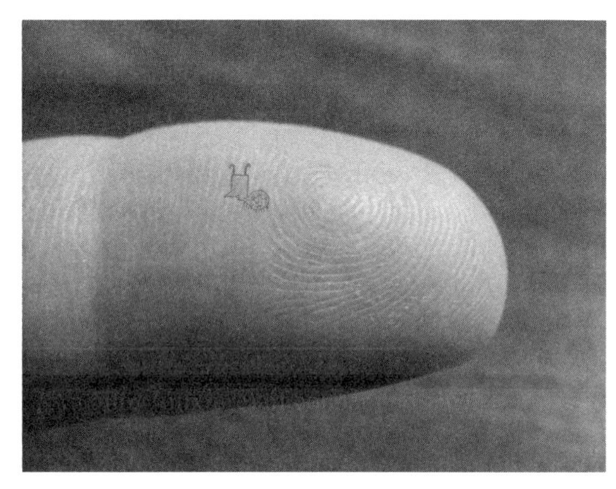

To Fred, it was the most obvious thing in the world.

To Kitty, with her 24-year-old eyeballs, nothing was there.

She changed the subject. "It seems a shame to me that they waste all that water in that fountain. They squirt it up, it flows over the waterfall, and then probably heads into the Great Lake."

"The water isn't wasted," Fred explained. He drew a picture of how the water goes around in a circle. The same water is used over and over again.

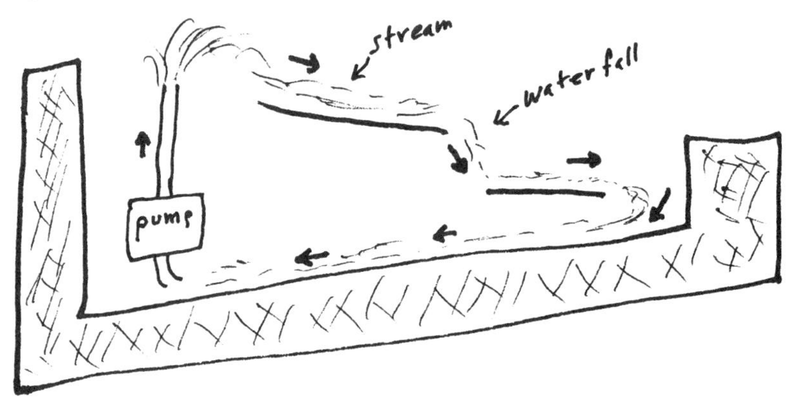

Fred was so excited that he said the water went in a *circle*. Everyone knows that this

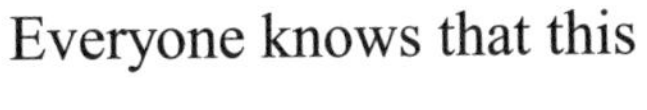

isn't a circle. It is a loop.

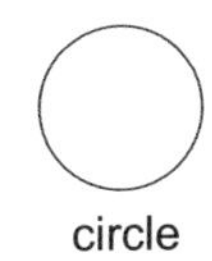

The water makes a circuit. (SIR-kit) A circuit is a journey that goes around and comes back to where it started.

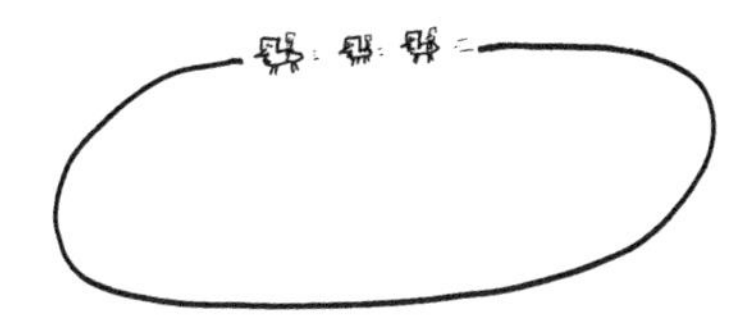

When horses go around a race track, they make a circuit.

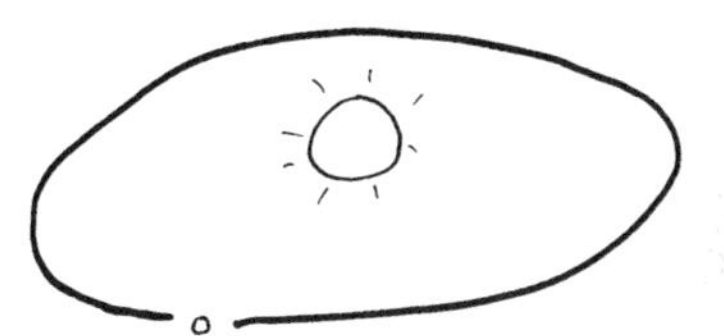

When the earth goes around the sun, it makes a circuit.

One of the most famous circuits is the **electric** circuit. A battery pumps electrons. They run around the wire, go through the lightbulb, and head back to the battery.

Wait a moment! I, your reader, have a couple of questions.

Okay. What do you have in mind?

First of all, what is an electron? Does that have to do with voting?

No, that's an election. An electron is—do you want the hundred-year-old, super-easy definition or do you want today's conception of an electron which may involve second-order partial differential equations and a discussion of wave-particle duality?

Does "second order" mean something like, "I'd like another helping of potatoes"?

Not quite. Here's an example of a second-order partial differential equation: $\frac{\partial^2 x}{\partial t^2} + \sin x \frac{\partial^2 y}{\partial t^2} = \sqrt{y} \log t$.

I'm not even going to ask what ∂ means! Give me the elevator definition of an electron—a quick definition that you could give me if we were riding in an elevator together.

An "elevator definition"? I've never heard of that before. Did you make that up?

Please. An easy definition of electron.*

Okay. Everything is made of atoms. Oops! That's an elliptical construction. I left out an important word. I should have said, "Every *physical* thing is made of atoms." Truth, love, and the number 398 don't contain atoms.

In the late 1800s, they were still debating whether things were made of atoms. One of the early models of what atoms looked like had electrons flying around a bunch of protons.

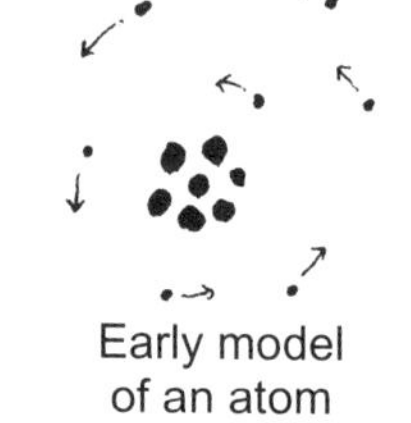

Early model of an atom

The electrons have a negative charge. The protons have a positive charge. The protons remind me of people huddling together while the electrons remind me of birds flying overhead.

The birds—I mean the electrons—are easily detached from the atom. They fly through the wires of a circuit. They come out of the negative (–) end of the battery, go through the lightbulb, and come back to the positive end of the battery.

---

* This is not a complete sentence. It is called an elliptical construction. In everyday speech, we often use elliptical constructions.

If you say, "I can draw prettier pictures than you," you don't mean that your pictures are prettier than my face. You mean, "I can draw prettier pictures than you *can draw*."

If the reader had not used an elliptical construction, the sentence might have been, ***Please give me an easy definition of the word electron.***

In formal English (such as school essays), fill out all the constructions and make every sentence complete.

In everyday English (emails to friends or talking to your dog), you may use elliptical constructions freely.

You said you had a couple of questions.

Yes. What if I cut the wire. How are those little electrons going to make their circuit?

They aren't.* If we start with

and cut the wire, the electrons can't move.

How can you tell?

Easy. The lightbulb goes out.

Your Turn to Play

1. Looking at the bill on the first page of this chapter, compute what percent of the GL&L bill was the strawberry milkshake.

2. Adding up the GL&L bill was easy.

$$\begin{array}{r} 3.00 \\ 2.50 \\ +\ 2.75 \\ \hline 8.25 \end{array}$$

Add up these numbers: 4.85 + 1.6 + 3.111

---

* Another elliptical construction. Filling out the construction: They aren't going to make their circuit.

.......COMPLETE SOLUTIONS.......

1. We want to know what percent of 8.25 is 2.75.

?% of 8.25 is 2.75

We don't know both sides of the *of*, so we divide the number closest to the *of* into the other number.

$$8.25\overline{)\,2.75}$$

In this case, this is the hard way to do the problem.

$$\begin{array}{r} 0.33\,\frac{275}{825} \\ 825.\overline{)\,275.00} \\ \underline{247\,5} \\ 27\,50 \\ \underline{24\,75} \\ 2\,75 \end{array}$$

Let's stop and do it an easier way. We want to divide 2.75 by 8.25.

$\frac{2.75}{8.25}$ = [multiplying top and bottom by 100] = $\frac{275}{825} = \frac{11}{33} = \frac{1}{3}$

From the Nine Conversions that you memorized in *Life of Fred: Decimals and Percents*, one-third is equal to $33\frac{1}{3}\%$.

The strawberry milkshake was $33\frac{1}{3}\%$ of the total bill.

2. To add up 4.85 + 1.6 + 3.111 we need to line up the decimals.

$$\begin{array}{r} 4.85 \\ 1.6 \\ +\ 3.111 \\ \hline 9.561 \end{array}$$

The whole point of this exercise is lost if you used a calculator.

Chapter Thirty-two
A Small History

As Fred and Kitty stood near the fountain, Fred noticed a big flashing sign over the GL&L Café building. Kitty asked Fred if he was hungry. Fred shook his head.

Kitty said, "Hey! They misspelled *baloney.*"

Fred explained to her that *baloney* is an American word that entered the language near the end of World War I (1918). *Baloney* is slang for nonsense. People will say, "That's a bunch of baloney."

Bologna is a cooked and smoked sausage. *Bologna* and *baloney* are both pronounced as beh-LOW-nee.*

"Why is bologna spelled so weirdly?" Kitty asked.

Fred had a story to tell. It was about electrical circuits.

Once upon a time about a couple hundred years ago (1781**), there lived a man named Luigi Galvani. (lew-EE-gee gal-VAH-nee)

He lived in the city of Bologna in the northern part of Italy. He worked at the University of Bologna.

---

* English is rarely simple. Some bologna makers will label their product as "baloney."

** Five years after the signing of the Declaration of Independence.
1776 + 5 = 1781

Kitty interrupted: "Can you imagine going to a school called baloney? That would be embarrassing."

Fred continued.

In Italy, Bologna is pronounced beh-LOW-nya. The "g" is silent.

Luigi was playing with frogs at the university.

Kitty: "This guy Louie seemed really strange. An adult playing with frogs?" [elliptical construction]

His name was Luigi Galvani.

Kitty: "Yeah, but I bet his mom, Mrs. Galvani, called him Louie."

They were dead frogs. Luigi was dissecting them as a biology experiment.

Kitty: "Yuck! Cutting up dead frogs. I would rather be playing golf."

And I would rather be a mathematician, but let me continue my story about electrical circuits.

The dead frog was held down with brass* hooks. His steel scalpel (medical word for knife) touched the frog's leg and the brass hook at the same time.

The frog leg moved . . . jumped . . . twitched.

dead frog

Kitty: "I bet Louie almost fell off his chair. He thought the stupid frog was dead. Ha. Ha. Big surprise for him."

The big surprise was that the frog was dead—and its leg moved. This galvanized** Galvani into ten years of research on what he called "animal electricity."

He had created an electrical circuit. After a decade of research, he published in 1791 an announcement that there was "animal electrical fluid" and battery-like things in frogs that supply the energy.

A year later, Alessandro Volta, working at the University of Pavia (200 km away), said, in effect, that Galvani was all wet with his animal

---

* An alloy (mixture) of copper and zinc.

** To galvanize is to startle into sudden activity. The word *galvanize* entered the English language around 1800. Luigi Galvani was working in the late 1700s. Alert readers may start to see a connection.

electrical fluid idea. The battery wasn't in the frog; it was in the brass-steel connection—where the steel scalpel touched the brass hook.

So Volta made batteries. He experimented using different pairs of metals to see which ones worked best.

That is why, today, we put batteries instead of frogs in our flashlights.

Kitty clapped her hands. She liked Fred's History of Circuits.

Your Turn to Play

1. Today, we put 1.5-volt batteries in our flashlights. Guess what famous man *volts* was named after.

2. Bologna and Pavia are 200 km apart. Using a conversion factor, compute how far that is in miles. (One kilometer equals about 0.6 miles.)

3. Nerves control the movement of muscles. Your brain sends a signal to the nerves in your arm: RAISE THAT PIECE OF PIZZA TO THE MOUTH. The muscles obey.

When Galvani made the frog's leg jump, he created a circuit. The two metals (brass/steel) made a battery. The electrons went from this battery through the scalpel to a nerve in the frog's leg. Then they traveled to the body of the frog and then back to the brass hook.

tweet!

Birds sitting on high voltage power lines are not shocked. Why not?

.......COMPLETE SOLUTIONS.......

1. If you guessed that volts was named after George Washington, you would get a grade of N.C. (Not Close)

If you guessed Luigi Galvani, then you could claim that you read part of the chapter.

If you said that volts was named after Alessandro Volta, you get an A.

2. Since 1 km equals 0.6 miles, the conversion factor will either be

$$\frac{1 \text{ km}}{0.6 \text{ miles}} \text{ or it will be } \frac{0.6 \text{ miles}}{1 \text{ km}}$$

$$\frac{200 \text{ km}}{1} \times \frac{0.6 \text{ miles}}{1 \text{ km}} = \frac{200 \cancel{\text{km}}}{1} \times \frac{0.6 \text{ miles}}{1 \cancel{\text{km}}} = 120 \text{ miles}$$

3. Electricity flows only when there is a closed circuit—a loop. Electrons flowing into the bird's feet would have nowhere to go.

If, on the other hand, the bird cut the power line and held each end with one of its wings, then the current would flow through the bird.

History can sometimes be loosey-goosey and squishy. Some historians say that Galvani touched the frog's nerve with a steel scalpel that was also touching the brass hook that was holding down the frog.

Another source talks about his hanging animals up on an iron railing using copper hooks and observing the twitching.

In another source, it was Galvani's assistant—and not Galvani—who first made the frog's leg move.

Can you think of an event in history where it is really important to know what actually happened?

Chapter Thirty-three
Kitty Café

Kitty looked back at the flashing Lunch Special Bologna Sandwiches sign. Fred could not imagine what she was thinking. It hadn't been 15 minutes ago that she had wolfed down a hamburger, onion rings, and a strawberry milkshake. He waited for her to speak.

"We need to go back to the café," she said.

She walked and Fred jogged to stay up with her.

She found the waiter who had served her breakfast.

He said, "Hi. I'm Michaelanthony. How may I be of service?"

Fred noticed that he looked a lot like that famous pizza guy Stanthony. He remained quiet while the adults talked.

Kitty said, "I have a couple of questions."

Michaelanthony nodded.

"First, I would like to know if your bologna sandwiches are as good as your hamburgers."

"Ma'am, they are even better. I specially import them from the city of Bologna."

"Yes. I know about that city. That's where Luigi Galvani made his great electrical circuit experiments."

Michaelanthony smiled.

Fred thought to himself Heavens! She's going to order another meal. Little beads of sweat started to form on Fred's forehead.

"I would like to speak to the owner of GL&L, if I may," she said.

"If you have a complaint, I would be glad to help you," Michaelanthony said. "I am the owner of this café."

"Perfect! I want to buy your café. I'll pay you whatever you like." She held out eight one-hundred dollar bills.

Mike didn't say anything. He took the money, handed her his apron, and walked away.

She turned to Fred and said, "You are too young to be in school yet. You have some free time. Would you like to help out a bit?" She handed him a bucket of black paint and a brush. She told him to think of a new name for the GL&L Café.

He thought of ☞ Kitty's Kitchen
☞ Kitty Cat Cooking
☞ Kitty Cuisine
☞ Kitty Litter (He was being silly.)
☞ Kitty and Mouse
☞ Kitty by the Lake
☞ Eat Here

He was starting to run out of decent ideas. He finally settled on Kitty Cafe and repainted the front of the building.

The letters weren't exactly straight, but that was the best that Fred could do.

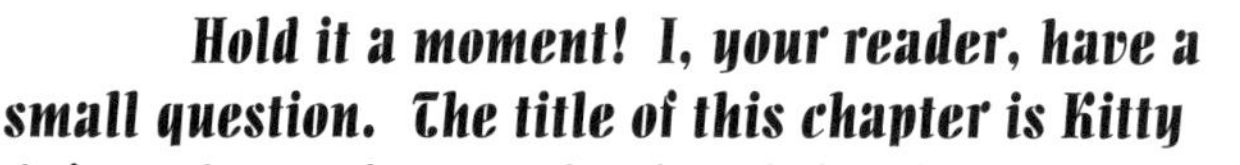

Hold it a moment! I, your reader, have a small question. The title of this chapter is Kitty Café and yet when Fred painted the sign, he wrote Kitty Cafe and didn't use the ´ mark over the e. Why?

In English you can write it either way—café or cafe. Both are correct. I prefer café and used that as the chapter title. Fred liked cafe and used that. We borrowed the word café from the French in the 1780s—the very same decade* that Galvani was making dead frogs twitch.

Thank you. I'm assuming that those two things happening in the same decade—borrowing café and Galvani's experiments—are just a coincidence.

Yes. Galvani was in Bologna, Italy. France was a zillion miles away from Italy. They didn't have automobiles back then.

In French, *café* means coffee. However, Kitty was speaking in English, so when she said that she wanted a café, Michaelanthony handed her the store and not just a cup of black liquid.

Now you, my reader, can speak (a little) French.

---

* Decade = ten years.

Kitty looked at Fred's sign. She saw that he was good with ideas, but he painted like a five-year-old.

She smiled and asked him, "What shall we put you in charge of?"*

Fred thought to himself In charge of? I don't understand. This woman hit me in the head with a golf ball. I dove into the lake to retrieve the ball for her. She took me out to breakfast. Now I'm supposed to help her run a cafe that she bought? On the other hand, this sounds like fun.

Fred answered, "Whatever. [another elliptical construction] I'm willing to learn any part of the business."

All that Kitty knew about Fred was that he was five years old, he could swim, he had a nice personality, he didn't eat much, and he knew a lot about the history of electrical circuits. She didn't know that he was a world-famous professor of mathematics at KITTENS.

"I am going to appoint you as chief electrician for Kitty Cafe," she announced.

Chief electrician! Fred thought. All I know is Galvani made an electrical circuit with his steel scalpel, the brass hooks, and the poor dead frog. Galvani thought it was "animal electrical fluid," but Volta corrected him by showing the electricity came from the steel/brass battery he had made. That's all I know.

"I would be glad to be your chief electrician," Fred answered. "But first I need to be excused for about ten minutes."

"Of course," Kitty said.

What Kitty thought Fred needed to be excused for was different than what Fred had in mind.

---

* It's important to know where to put the question mark at the end of a quotation.

In English, you sometimes write ?" and sometimes write "?, depending on the circumstances.

First case: If you are quoting someone who is asking a question, then you write ?" *Kitty asked Fred, "What shall we put you in charge of?"*

Second case: If the quoted material, itself, is not a question. *Was it Jefferson who said, "A slave is someone who does not enjoy the fruits of his own labor"?*

Fred ran across the Great Lawn, past the Great Lake, through the Great Woods,

One entrance to the Great Woods

near the construction site for the Feynman Physics Building,

through the rose gardens, past the university chapel and the tennis courts, to his building,

up two flights of stairs, down the hallway past the nine vending machines (four on one side and five on the other),

through his doorway that didn't have a door, around the safe,

and to his walls of books. He pulled his copy of Prof. Eldwood's *Elementary Electric Circuits* off the shelf.

The first thing that Fred noticed was the funny way that electrical circuits are drawn.

Instead of everyone draws

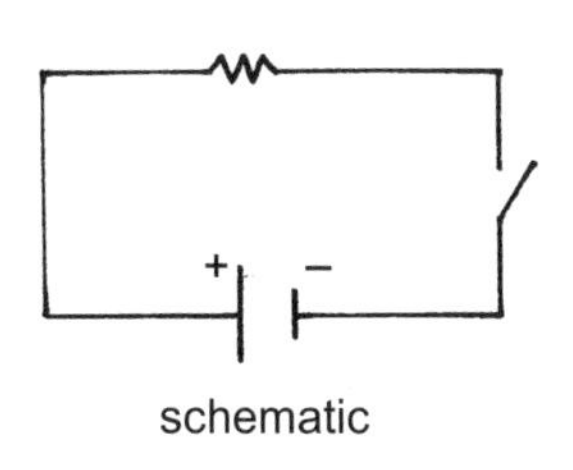

schematic

This is called a schematic (skee-MAT-ic) representation, a schematic drawing, a circuit diagram, an electrical diagram, or simply a schematic. Schematics are used by electricians everywhere in the world. They are simplified standard symbols.

The jagged line —WW— represents any kind of resistance. It could be a lightbulb, a hair dryer, a toaster, or a television set.

Your Turn to Play

1. Most people don't usually carry eight one-hundred dollar bills around with them. Kitty was an exception. She played golf for fun, but her real joy was in owning businesses. Years ago, she started out by running a lumber store. She called it Kitty Wood.

Then she owned a pet store, which she named Kitty. She has owned many businesses over the years. Some have brought in a lot of money; some have gone bankrupt.*

After each business gets started, she hires and trains a manager for that business. That gives her time to play golf and buy other businesses.

Each month, she earns $30,000. Using conversion factors, how much does she make each hour? Assume 30 days in a month and 8 hours in a workday.

2. Eight hundred dollars is what percent of $1,000?

3. What symbol is used by electricians to indicate an electric pizza oven in a circuit diagram?

* With two independent clauses, I had three choices.

✏ **Make two sentences:** Some have brought in a lot of money. Some have gone bankrupt.

✏ **Use a conjunction:** Some have brought in a lot of money, and some have gone bankrupt.

✏ **Use a semi-colon:** Some have brought in a lot of money; some have gone bankrupt.

.......COMPLETE SOLUTIONS.......

1. The conversion factor for changing between months and days is either $\frac{\text{1 month}}{\text{30 days}}$ or it is $\frac{\text{30 days}}{\text{1 month}}$ (Many businesses, such as gas stations, are open seven days a week. There are people who work on Sundays, such as clergymen.)

The conversion factor for changing between days and hours is either

$\frac{\text{1 day}}{\text{8 hours}}$ or it is $\frac{\text{8 hours}}{\text{1 day}}$

$$\frac{\$30{,}000}{\text{month}} \times \frac{\text{1 month}}{\text{30 days}} \times \frac{\text{1 day}}{\text{8 hours}}$$

$$= \frac{\overset{1{,}000}{\$\cancel{30{,}000}}}{\cancel{\text{month}}} \times \frac{1\ \cancel{\text{month}}}{\cancel{30}\ \cancel{\text{days}}} \times \frac{1\ \cancel{\text{day}}}{\text{8 hours}}$$

$$= \frac{\$1{,}000}{\text{8 hours}} = \frac{\$125}{\text{hour}}$$

$$\begin{array}{r} 125 \\ 8\overline{)1000} \\ \underline{8} \\ 20 \\ \underline{16} \\ 40 \\ \underline{40} \end{array}$$

2. 800 is ?% of 1000

We don't know both sides of the *of*, so we divide the number closest to the *of* into the other number.

$$\begin{array}{r} 0.8 \\ 1000\overline{)800.0} \\ \underline{8000} \end{array} \qquad 0.8 = 0.80 = 80.\% = 80\%$$

Eight hundred dollars is 80% of a day's earnings for Kitty.

3. Electricians would use the symbol —\/\/\/— .

I would use the symbol —☺— .

Making Tons of Money

Most millionaires in the United States are owners of small businesses. They are not baseball players, actors, or rock stars. The difference is that you read about the baseball player, actor, or rock star who makes big money, and you don't read about your neighbor down the street who owns a toy store. Your neighbor doesn't want the publicity.

Very, very few baseball players, actors, or musicians make lots of money.

The Rule: Having a good marriage will make you happier than having a ton of money in a safe. Of course, having both is also nice.

Chapter Thirty-four
Volts and Amperes

Fred continued reading in his Prof. Eldwood's *Elementary Electric Circuits* book. He had a lot to learn if he was going to be the chief electrician for the Kitty Cafe.

Eldwood started out with a circuit diagram in which the switch was closed. He said that the battery was like a heart that pumped the electrons around the loop.

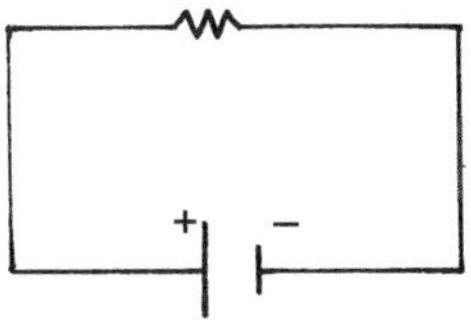

Eldwood wrote: In a *Baby Electric Circuits* book, you might tell the little three-year-olds that the battery makes electrons and spits them out the negative terminal of the battery. Then they run around, go through the resistor, and come back to the positive terminal of the battery.

That is poppycock!* Every atom in the wire, every atom in the resistor, and every atom in the battery has electrons. A heart doesn't make blood; it pumps it. A battery doesn't make electrons; it makes the current flow.

Fred was glad he wasn't reading a *Baby Electric Circuits* book. He wanted the truth.

How hard the battery shoves the electrons is measured in **volts**. A 6-volt battery shoves electrons twice as hard as a 3-volt battery.

How fast the current flows is measured in **amperes**. A wire that is carrying 6 amperes (or 6 amps or 6 A) is moving twice as many electrons in the same amount of time as a wire carrying 3 amperes.

If you wanted to measure how fast the electrons are running through a particular place in a wire, you just stick in an ammeter. (No *p* in that word! Did I ever mention that English is harder than math?)

---

* Poppycock means nonsense. Prof. Eldwood is a very old man and he sometimes uses words that show his age. Some teenagers use slang to show how young they are.

The symbol for an ammeter is —Ⓐ—

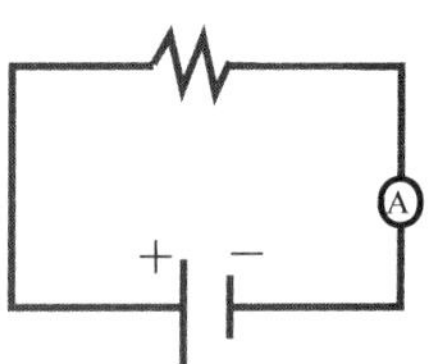

Suppose you had a circuit that looked like this: and the ammeter read 30 A.

If you moved the ammeter to a different spot on the circuit, make a guess what the ammeter would read.

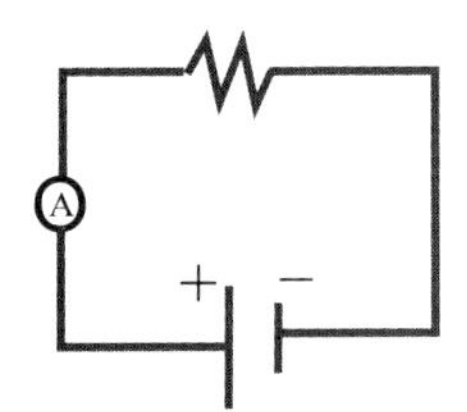

If you guessed 30 A, you were right. The electrons don't disappear as they speed around the circuit loop.

One ampere represents 6.24×10^{18} electrons per second.

10^{18} means eighteen 10s multiplied together.

$6.24 \times 10^{18} = 6.24 \times 10 \times 10 \times 10 \times 10 \times 10 \times 10 \times 10 \times 10 \times 10 \times 10 \times 10 \times 10 \times 10 \times 10 \times 10 \times 10 \times 10 \times 10$.

When you multiply a number containing a decimal by ten, you move the decimal over one place to the right.

$6.24 \times 10^{18} =$ 6,240,000,000,000,000,000, which is six quintillion, two hundred forty quadrillion if you prefer English.

The whole point is that if an ammeter is reading 1 A, it means that 6,240,000,000,000,000,000 electrons per second are flowing through that point in the wire, which means that electrons must be very tiny little things.

You can't see electrons with a microscope.

In the second chapter of Prof. Eldwood's book, he wrote: The fun begins when you have two resistors in the circuit. If ammeter #1 reads 8 A, then what will ammeter #2 read?

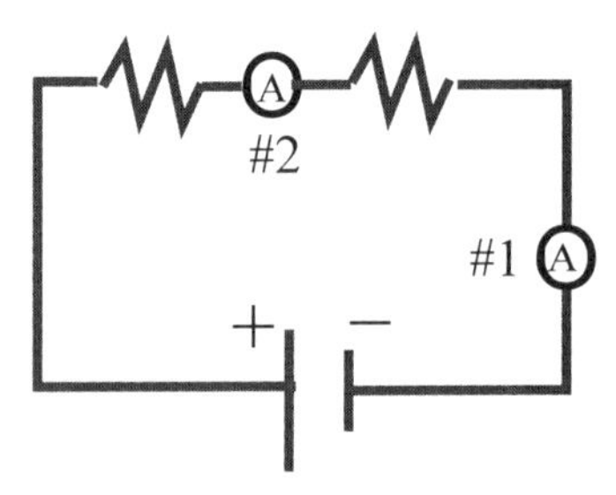

The answer is that ammeter #2 must also read 8 A.

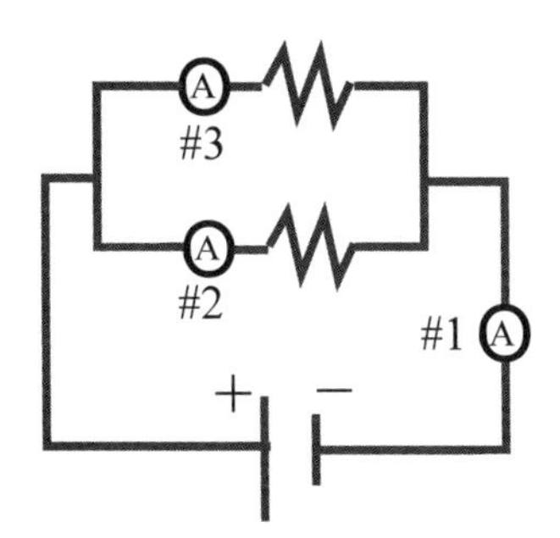

Since this isn't a *Baby Electric Circuits* book, I can ask a harder question. Suppose that ammeter #1 reads 8 A and ammeter #2 reads 5 A. What will ammeter #3 read?

Fred looked at that diagram for a minute and thought *If the flow through ammeter #1 is 8 A and if 5 A is flowing through ammeter #2, then 3 A must be flowing through ammeter #3.*

Your Turn to Play

1. Ammeter #2 reads 14 A. Ammeter #3 reads 6 A. What do ammeters #1 and #4 read?

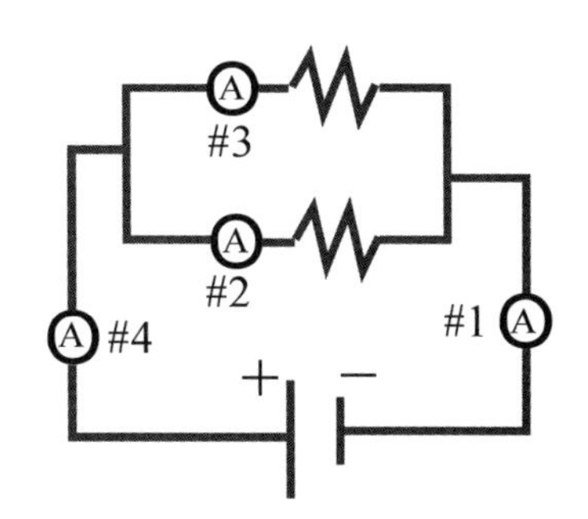

2. In some simple circuit, suppose a 1.5-volt battery produced a current of 3 amperes. If you replaced that battery with a 5-volt battery, how much current would be produced? (Use a conversion factor.)

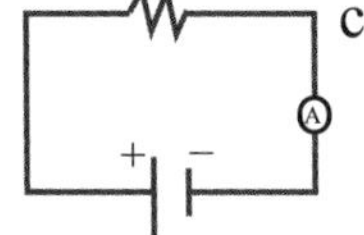

3. How many electrons per second are flowing through a ⅓ amp circuit?

.......COMPLETE SOLUTIONS.......

1. 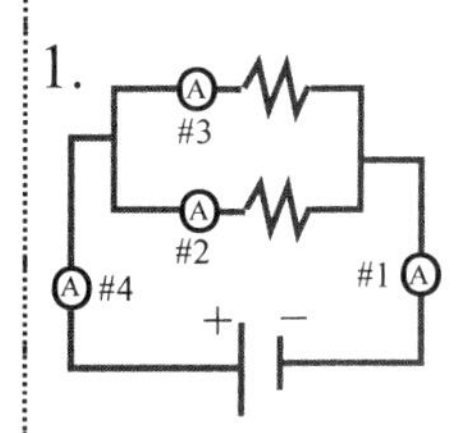

If #2 reads 14 A and #3 reads 6 A, then a total of 20 A must be running through that pair. Therefore, both #1 and #4 will read 20 A.

2. 1.5 V matches up with 3 A. The conversion factor will either be

$\frac{1.5\text{ V}}{3\text{ A}}$ or it will be $\frac{3\text{ A}}{1.5\text{ V}}$

$$\frac{5\text{ V}}{1} \times \frac{3\text{ A}}{1.5\text{ V}} = \frac{5\cancel{\text{V}}}{1} \times \frac{3\text{ A}}{1.5\cancel{\text{V}}} = \frac{15\text{ A}}{1.5} = 10\text{ A}$$

$1.5\overline{)\,15}$ becomes $15.\overline{)150.}$ = 10.

3. Using a conversion factor, $\frac{⅓\text{ A}}{1} \times \frac{6.24 \times 10^{18}\text{ electrons}}{1\text{ A}}$

$$= \frac{⅓\cancel{\text{A}}}{1} \times \frac{6.24 \times 10^{18}\text{ electrons}}{1\cancel{\text{A}}}$$

$$= 2.08 \times 10^{18}\text{ electrons}$$

Or, if you prefer, $\frac{⅓\text{ A}}{1} \times \frac{6{,}240{,}000{,}000{,}000{,}000{,}000\text{ electrons}}{1\text{ A}}$

$$= \frac{⅓\cancel{\text{A}}}{1} \times \frac{6{,}240{,}000{,}000{,}000{,}000{,}000\text{ electrons}}{1\cancel{\text{A}}}$$

$$= 2{,}080{,}000{,}000{,}000{,}000{,}000\text{ electrons}$$

⅓ of 6.24 ⇛ $3\overline{)\,6.24}$ = 2.08

Chapter Thirty-five
Ohms

Fred had told Kitty that he would be back in about ten minutes. Already, four minutes had passed. Prof. Eldwood's book had taught Fred that how hard a battery shoves electrons is measured in volts, and how fast the current flows is measured in amperes.

If you double the volts, you will double the amperes.*

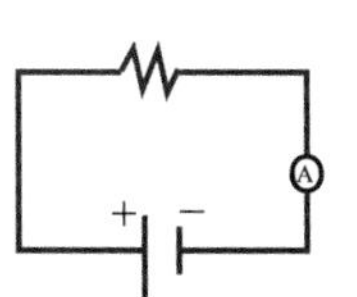

The only thing left to look at was the resistor, which could be a lightbulb, a hair dryer, a toaster, or a television.

Resistors slow down the flow of electrons. Have you ever noticed that lightbulbs, hair dryers, toasters, and televisions get warm when they are operating?

Friction turns motion into heat.

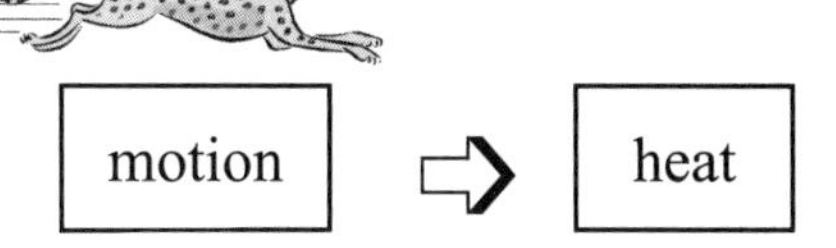

Resistors turn electricity into heat.

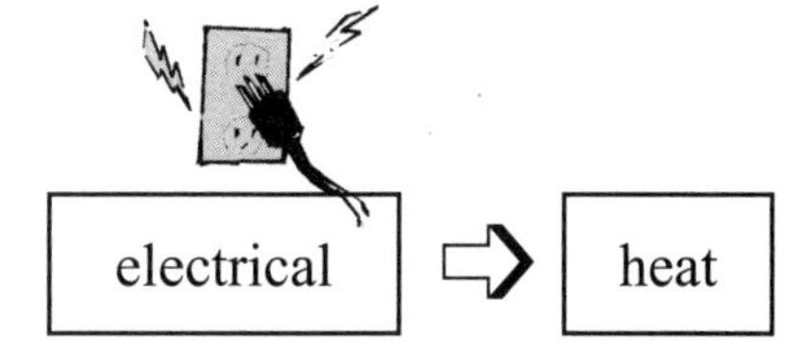

If you double the resistance, you will cut the flow of electrons in half.

If you are watering your garden and you get a kink in your hose, the flow is decreased.

---

* The fancy way of saying that is that the amperes are proportional to the volts. If you triple the volts, you will triple the amperes.

The easy part:

The pressure is measured in volts.
The flow of electrons is measured in amperes.
The resistance is measured in ohms.

The medium hard part:

Six volts is written as 6 V.
Six amperes is written as 6 A.
Six ohms is written as 6 Ω.

And now, before you, my reader, can have a chance to interrupt . . .

I was just about to ask about that Ω thing. It looks like a pair of twos that are kissing.

It has nothing to do with twos or with kissing. Ω is the Greek letter omega. We couldn't let O be used for ohms. Writing six ohms as 6 O would be asking for trouble. 6 O looks too much like 60 (sixty).

Why did they choose a Greek letter? I would have chosen R instead.

It was probably because of poetry.

Poetry! You gotta be kidding!

Omega sounds almost like ohms. They almost rhyme.*

Okay. If that's the hardest part of electric circuits, I can handle it. You taught me French (café = coffee) and now you teach me Greek.

Now comes the nutty part. When we do the math for electric circuits:

V stands for the voltage.	6 V
R stands for the resistance.	6 Ω
I stands for the rate of flow of the current.	6 A

"I" for current! Why not A for amps? Why not C for current? Why not F for flow. Using "I" is totally crazy.

I told you so.

---

* Ω Some people say oh-MAY-geh and some say oh-ME-geh.

It is tradition. Everybody uses I for current.

Tell me why the top row of a keyboard is QWERTYUIOP and not ABCDEFGHIJ.* Tradition.

Tell me why you turn screws to the right to tighten them. Tradition.

Tell me why you read from left to right? Tradition.

We get so used to reading from left to right.

Tell me why we ask, "How are you?" when we don't expect a real answer.

Tell me why we wear clothes on days ending in "y."

Not all traditions are good, but changing a tradition is sometimes very hard.

We shall use I for current. . . . even though it is nutty.

Your Turn to Play

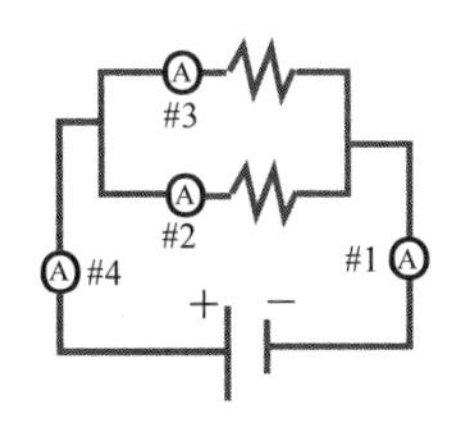

1. Ammeter #1 reads 60 A and ammeter #2 reads 5 A. What do ammeters #3 and #4 read?

2. Some traditions are fairly easy to change. In really old books, the word *today* was written as *to-day*. The gradual changeover was not too hard to make.

When you are driving, a red light means stop and a green light means go. What do you imagine would be the result if we changed the tradition and made red = go and green = stop?

---

* Originally, when typewriters (remember them?) were invented,
QWERTYUIOP
ASDFGHJKL
ZXCVBNM,.? was the arrangement selected so that there was less chance that the typewriter would jam up when two keys were hit at almost the same time. This is something that your grandparents might remember.

.......COMPLETE SOLUTIONS.......

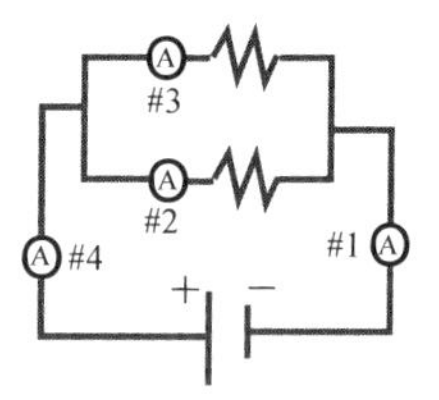

1. If the current through #1 is 60 A, then the current through #4 must also be 60 A.

The combined current through #2 and #3 must also be 60 A. If the current through #2 is 5 A, then 55 A must be passing through #3.

2. Tons of auto accidents.

If you have been driving for years and you see a green light, you think "go." You don't have to think about it. It would take months and lots of accidents for many drivers to reprogram their brains.

Some people might have to give up driving because they couldn't make the adjustment to the new rules.

There are lots of traditions that you hardly notice. Look at the front door to your house or apartment. Have you ever observed that the door opens inward?

Your friend knocks on your front door. You open the door and let him in. If the door opened outward, you would have to dance around him in order to let him in and shut the door.

Look at the front door to any business or other public building. It opens outward.

You and a hundred other people are in that building. There is a fire and everyone rushes to the door. If it only opened inward, then if the first person couldn't get the door opened quickly, there is the possibility that the crowd would crush against that door making it impossible to open. The crowd would be toast.

Chapter Thirty-six
Ohm's Law

Fred had three minutes left before he was supposed to get back to Kitty and become the chief electrician at the Kitty Cafe. Luckily, there was only one more chapter to read in Prof. Eldwood's *Elementary Electric Circuits* book. The chapter talked about how volts (V), current (I), and resistance (R) are related to each other.

Prof. Eldwood wrote: One day in 1787, Mrs. Ohm had a baby boy that she named Georg.

Wait a minute! I, your reader, found a typo. Prof. Eldwood wrote "Georg" instead of "George."

That's not an error. The Ohms lived in Germany. They think we are the weird ones since we put an *e* on the end of *Georg.*

You mean that they have a different tradition than we do?

Yup. Now let's get back to Prof. Eldwood's story.

When Georg grew up, he liked to play with electricity. He borrowed his father's 12 V car battery.

Hold it! This was around 1800. They didn't have cars back then.

Please, just let Prof. Eldwood tell his story. In his book, he labeled this story as apocryphal.

But you didn't tell me that.

Fred only has three minutes left. I'm leaving out the less important parts of Eldwood's book.

Okay.

So Georg took the 12 V battery, an ammeter, some wire, and various devices and made a few different circuits.

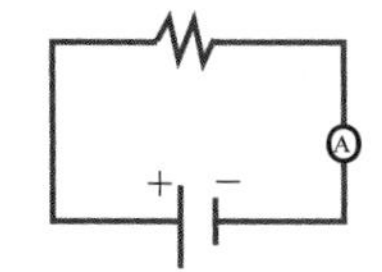

When Georg used a 1 Ω lightbulb, the ammeter read 12 A.
When he used a 2 Ω bell, the ammeter read 6 A.
When he used a 3 Ω radio, the ammeter read 4 A.
When he used a 4 Ω phonograph, the ammeter read 3 A.

Here is the chart that Georg Ohm made from his electrical experiments:

R	I	V
1	12	12
2	6	12
3	4	12
4	3	12
6	2	12
12	1	12

It didn't take Georg long to realize that resistance times current is equal to volts. RI = V.

Or, if you like, V = RI or V = IR or $\frac{V}{I} = R$ or $\frac{V}{R} = I$.

When he showed this to his mother, she said, "We are going to call this Ohm's Law." And Georg became famous. His law is in every modern physics book.

And those were the last words of Prof. Eldwood's book. Fred closed the book, put it back on the shelf in his office, said goodbye to Kingie, walked around the safe, down the hallway past the nine vending machines, and down two flights of stairs.

He ran past the tennis courts, the university chapel, through the rose gardens, near the construction site for the Feynman Physics Building, through the Great Woods, past the Great Lake, across the Great Lawn, and into the Kitty Cafe.

Kitty greeted him, "Welcome back, Chief Electrician."

"Thank you. I got everything taken care of. Now where can I start?"

Kitty handed him a giant flashlight and asked him to put batteries in it.

Fred found four batteries, lined them up, and put them into the flashlight.

He tested the bulb with an ohmmeter* and found that it had a resistance of 10 ohms.

Fred wanted to do a perfect job for Kitty so he also drew a circuit diagram.

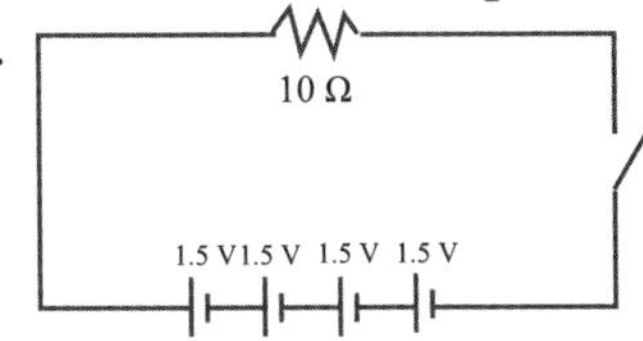

The batteries are in **series**—all in a row. When batteries are in a series you can just add the volts. $1.5 + 1.5 + 1.5 + 1.5 = 6$ volts

Ohm's Law can be written in many forms: $RI = V$ or $V = RI$ or $V = IR$ or $\frac{V}{I} = R$ or $\frac{V}{R} = I$. All you need to do is remember any one of the forms. Fred liked $V = IR$ because it reminded him of *virtue* (being good). **V=IR TUE.**

Your Turn to Play

1. Complete this sentence: Fred told to Kitty that when the switch was closed the flashlight would have a current (I) of ___?___ amps.

   Start with $V = IR$. You know that $V = 6$ and $R = 10$.

2. Resistors in series work the same way as batteries in series. Just add up the ohms.

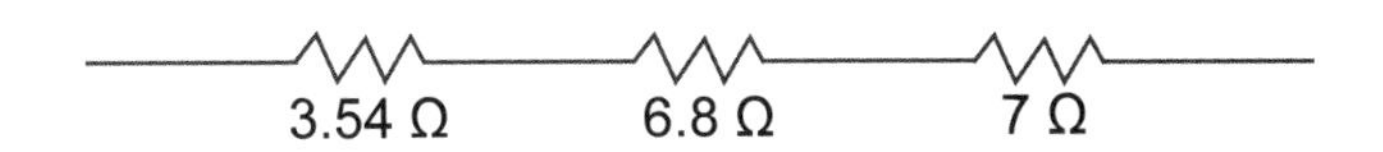

Complete this sentence: All these resistors can be replaced by a single resistor with ___?___ ohms.

---

* There really is such a thing as an ohmmeter. You can buy an ohmmeter—and a voltmeter—at electronics stores.

.......COMPLETE SOLUTIONS.......

1. V = IR becomes 6 = (I)(10).

In algebra, we often place the number in front of the letter.

6 = 10I

Time Out!

10I (ten eye) looks a lot like 101 (one hundred one), doesn't it?

The people who made the traditions for electric circuits were smart enough to use Ω for ohms instead of O.

However, they weren't smart enough to avoid using I for amperage (current flow). Someone should have told them that:

the letter I looks like the number 1	1I
the letter b looks like the number 6	6b
the letter O looks like the number 0	0O
the letter z can look like the number 2	2z

end of Time Out

To solve	$6 = 10I$
we divide both sides by 10	$\frac{6}{10} = \frac{10I}{10}$
and simplify	$0.6 = I$

The current through the flashlight is 0.6 amps.

2. To add numbers with decimals in them, line up the decimal points.

```
  3.54
  6.8
  7.
 -----
 17.34
```

All these resistors can be replaced by a single resistor with 17.34 ohms.

A small note to you, my reader. We have one more little bit of electricity to look at. I'm going to delay The Bridge for one more chapter. I hope you don't mind. I didn't think you would. [elliptical construction]

Chapter Thirty-seven
Parallel Circuits

Fred handed Kitty the flashlight filled with batteries, the schematic, and his computations showing that I = 0.6 amperes for the circuit. He made a small notation on his paper that the batteries were in series.

Fred was very proud of his work. His diagram showed that he had mastered all of Prof. Eldwood's *Elementary Electric Circuits*. His drawing was very neat.

Kitty took the flashlight and turned it on to make sure it was working. She threw Fred's paper in the garbage can.

A little tear came to Fred's eye.

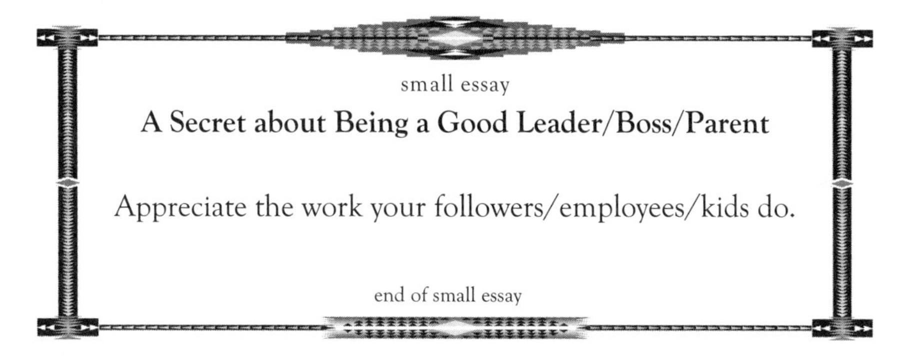

They headed off to the kitchen. She plugged in the mixer to see if it worked.

Fred mentally drew the schematic.

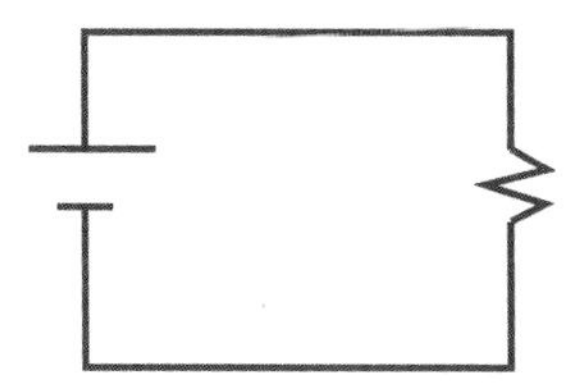

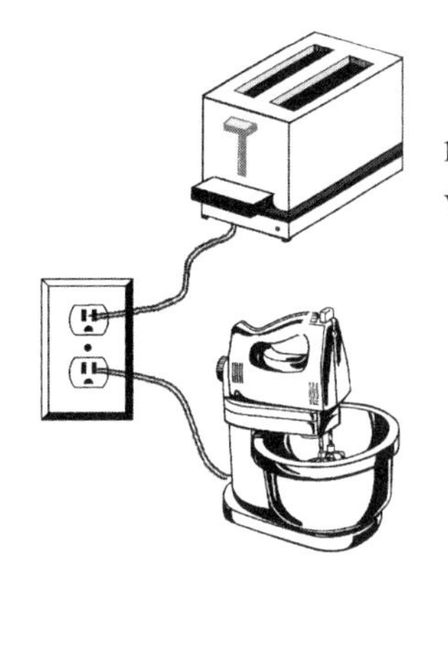

Then she plugged in the toaster. Both the mixer and the toaster were working. She was very pleased.

The schematic for this is not:

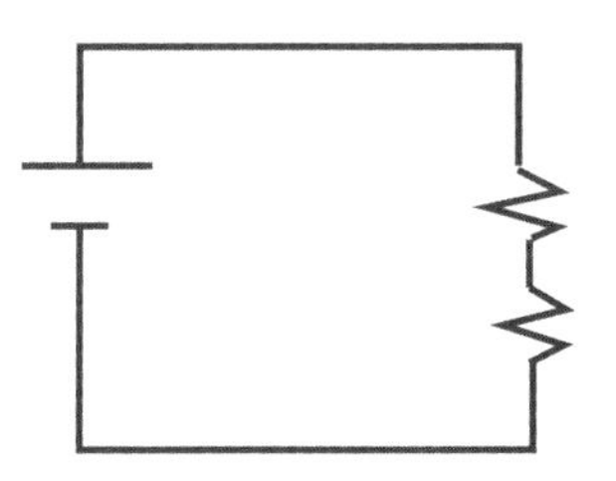

The mixer and toaster are not in series. If they were in series, things would be really bad. If you turned off the mixer, then the toaster would not work!

Instead, in house wiring, everything you plug in is in parallel.

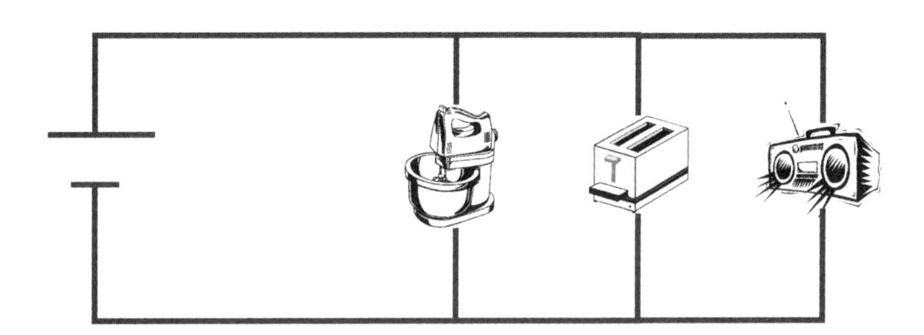

You don't have to turn on the mixer in order to play the radio.

If we stick in some ammeters, we can measure the current (I) through each part of the circuit.

I is the total current running through the circuit.

I_1 is the current running through the mixer.

I_2 is the current running through the toaster.

I_3 is the current running through the radio.

I counts electrons per second running through the battery.
I_1 counts electrons per second running through the mixer.
I_2 counts electrons per second running through the toaster.
I_3 counts electrons per second running through the radio.

$$I = I_1 + I_2 + I_3$$

The more appliances you turn on, the faster the current flows. If you don't believe me, turn on every appliance and every light in your house and go outside and look at your electric meter. It will be spinning like crazy.

The total electrons used each second is the sum of the electrons per second running through each of the appliances. (Actually, amperes is measuring 6,240,000,000,000,000,000 electrons per second, but let's keep things simple for a moment. Parallel circuits are hard enough. We don't want to be too fussy right now.)

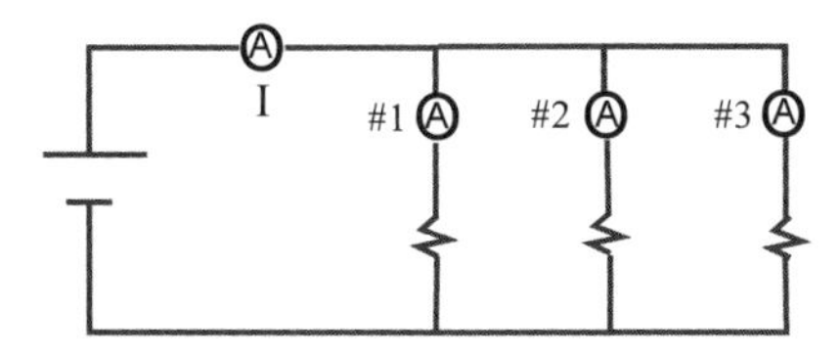

Think of it this way. Any electron passing through the ammeter marked I will have to pass through either #1, #2, or #3.

In any event, $I = I_1 + I_2 + I_3$. Now come two lines of math. Please fasten your seatbelts.

Line #1: $$\frac{V}{R} = \frac{V}{R_1} + \frac{V}{R_2} + \frac{V}{R_3}$$

Explanation: We used Ohm's Law. We replaced I by $\frac{V}{R}$

R is the resistance of the whole circuit. It is the resistance that the battery "feels."
R_1 is the resistance of the mixer.
R_2 is the resistance of the toaster.
R_3 is the resistance of the radio.
The voltage V is the same for all the appliances.

Line #2: $$\frac{1}{R} = \frac{1}{R_1} + \frac{1}{R_2} + \frac{1}{R_3}$$

Explanation: We divided every term by V. (Or, if you like, we multiplied every term by $\frac{1}{V}$)

For example, $\frac{V}{R} \times \frac{1}{V} = \frac{1}{R}$

Resistance
Series vs. Parallel

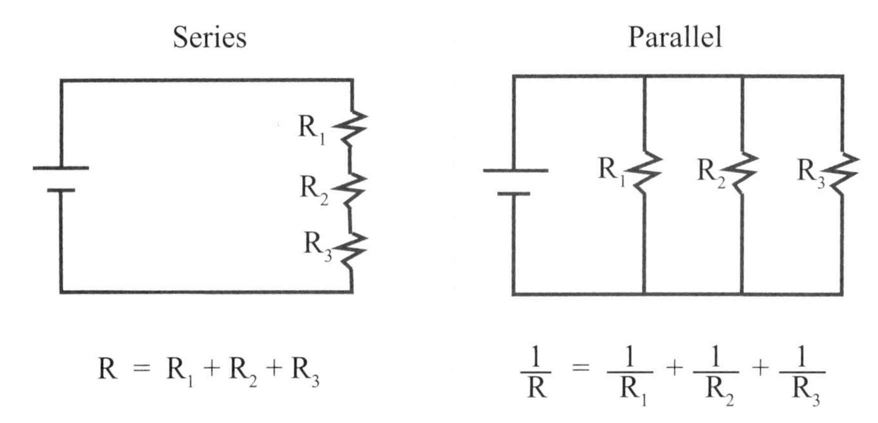

$R = R_1 + R_2 + R_3$

$$\frac{1}{R} = \frac{1}{R_1} + \frac{1}{R_2} + \frac{1}{R_3}$$

Kitty turned on the mixer, the toaster, and the radio.

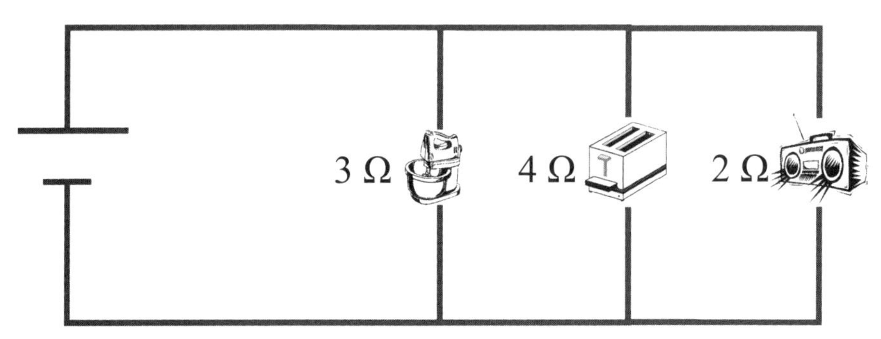

How to find R? R is the total resistance that the three appliances make.

$$\frac{1}{R} = \frac{1}{R_1} + \frac{1}{R_2} + \frac{1}{R_3}$$

$$\frac{1}{R} = \frac{1}{3} + \frac{1}{4} + \frac{1}{2}$$

Add the fractions $\frac{1}{R} = \frac{13}{12}$

Invert (flip upside down) $R = \frac{12}{13}$ The total resistance is $\frac{12}{13}\ \Omega$.

Small notes.

♪#1: One of the reasons you learned to add fractions (besides following recipes in the kitchen) was to be able to work with resistors in parallel.

♪#2: $\frac{1}{3} + \frac{1}{4} + \frac{1}{2} = \frac{4}{12} + \frac{3}{12} + \frac{6}{12} = \frac{13}{12}$

♪#3: If two fractions are equal to each other, such as $\frac{3}{4} = \frac{6}{8}$, then you can invert both of them and they will still be equal. $\frac{4}{3} = \frac{8}{6}$

We will prove that fact when we get to algebra.

Your Turn to Play

1. If two lightbulbs (4 ohms and 5 ohms) are in series, what is the total resistance?

2. (Part A) If two lightbulbs (4 ohms and 5 ohms) are in parallel, what is the total resistance?

(Part B) If those two lightbulbs are powered by an 18-volt battery, how many amperes are going through the battery?

(Part C) How many amperes are going through each bulb?

3. If two lightbulbs (0.1 ohms and 0.2 ohms) are in series, what is the total resistance?

4. If two lightbulbs (0.1 ohms and 0.2 ohms) are in parallel, what is the total resistance?

.......COMPLETE SOLUTIONS.......

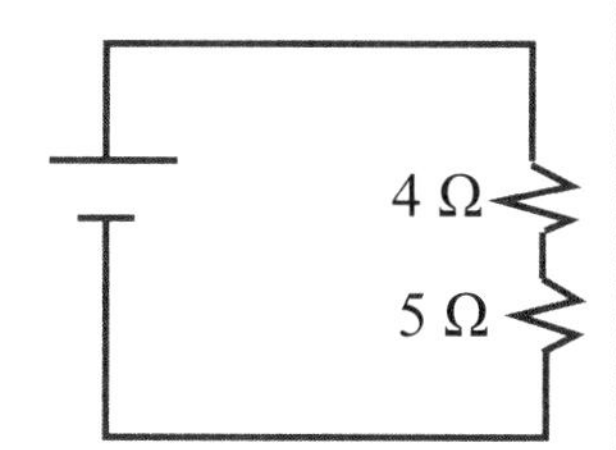

1. 9 Ω

When the resistors are in series, we may add the ohms. $4\ \Omega + 5\ \Omega = 9\ \Omega$

2. (Part A)

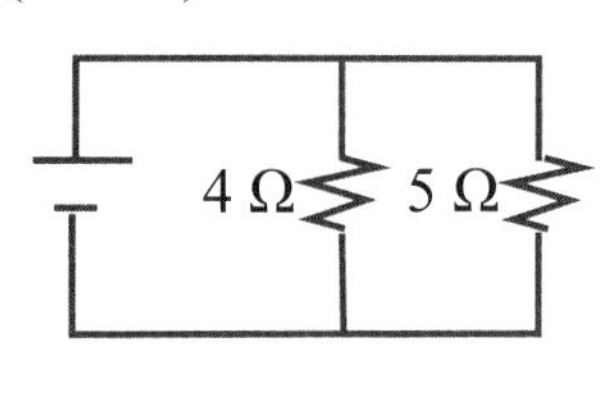

$$\frac{1}{R} = \frac{1}{R_1} + \frac{1}{R_2}$$

$$\frac{1}{R} = \frac{1}{4} + \frac{1}{5}$$

$$\frac{1}{R} = \frac{5}{20} + \frac{4}{20}$$

$$\frac{1}{R} = \frac{9}{20}$$

$$R = \frac{20}{9}\ \Omega = 2\frac{2}{9}\ \Omega$$

2. (Part B)

From Part A, we know $R = \frac{20}{9}\ \Omega$. We are given that V = 18 volts.

Ohm's Law: V = IR becomes $18 = I(\frac{20}{9})$

Multiply both sides by $\frac{9}{20}$ $\quad 18 \times \frac{9}{20} = I\,\frac{20}{9} \times \frac{9}{20}$

This trick may be new to you.

Simplify $\quad \frac{18 \times 9}{20} = I$

Do the arithmetic $\quad 8\frac{1}{10}$ amps $= I$

2. (Part C)

Ohm's Law: $\frac{V}{R} = I$ becomes $\frac{18}{4} = I$ for the 4 ohm bulb $= 4\frac{1}{2}$ A.

Similarly, $I = 3\frac{3}{5}$ amps through the 5-ohm bulb.

Note that $4\frac{1}{2} + 3\frac{3}{5}$ equals $8\frac{1}{10}$

The sum of the currents through each of the bulbs equals the current through the battery.

3. $0.1 + 0.2 = 0.3$ ohms

4. $\frac{1}{R} = \frac{1}{R_1} + \frac{1}{R_2} = \frac{1}{0.1} + \frac{1}{0.2} = 10 + 5 = 15 = \frac{15}{1}$

$R = \frac{1}{15}\ \Omega$

The Bridge

from Chapters 1–37

first try

Goal: Get 9 or more right and you cross the bridge.

1. Joe liked to get his boat and go fishing out on the Great Lake. He liked to say, "When I'm fishing, I can get away from everything." Unfortunately, Joe is easily bored, so he bought a long extension cord that he plugged into an outlet on a house on the beach. The other end of the cord was on his boat. He plugged his 50 ohm radio and his 60 ohm television set into the extension cord. When he turned them both on, what was the combined resistance?

[Joe had never heard of Thomas Merton, a world-famous author of more than 70 books on spirituality, including his best-selling autobiography, The Seven Storey Mountain. In December 1968 he gave a talk at an interfaith conference at a Red Cross Conference Center in Thailand. In his cabin he touched a faulty electric fan while he was still wet from bathing. He was found dead with the fan across his body. He was 53 years old. Electricity and water can be a fatal combination.]

2. Joe's shirt was very hard to button. It was the same shirt he had when he was eight years old. When it was buttoned, it pressed on his chest with a pressure of 5 pounds per square inch. How many pounds per square foot would that be?

3. In the chart on page 206, Georg Ohm found that when the voltage was held constant, he obtained the points (1, 12), (2, 6), (3, 4), (4, 3), (6, 2), and (12, 1). Plot those six points on a graph.

4. Joe shoved the radio over with his foot so that he could watch the television. It took 3.2 pounds of force to get his 4.8 pound radio moving. What is the coefficient of static friction between the radio and the boat?

5. Joe pushed a little too hard. His radio became unplugged and fell over the edge of the boat. The radio weighed 4.8 pounds. The density of water in the Great Lake is 62 $lbs./ft^3$. What additional piece of information do we need in order to know whether the radio sank or floated?

The Bridge

from Chapters 1–37

6. His radio was not the first thing that Joe lost while fishing. Of the 600 items that Joe has taken on his boat over the years, he has lost 500 of them overboard. What percent has he lost? (Round your answer to the nearest percent.)

7. Joe thought that the Law of Conservation of Energy meant that he should conserve his energy and not work too hard. In physics, the Law of Conservation of Energy doesn't mean laziness. What does it mean?

8. Joe put a board on his lap. That made a nice table so that he didn't have to keep holding his sandwich between bites. He liked to save as much energy as possible.

Darlene once told him that he was so lazy that if he could get someone else to chew his food for him, he would.

Joe didn't know whether he liked that idea. It gave him something to think about.

A wave came and started to tip the boat. This is a picture of when the sandwich first began to slide.

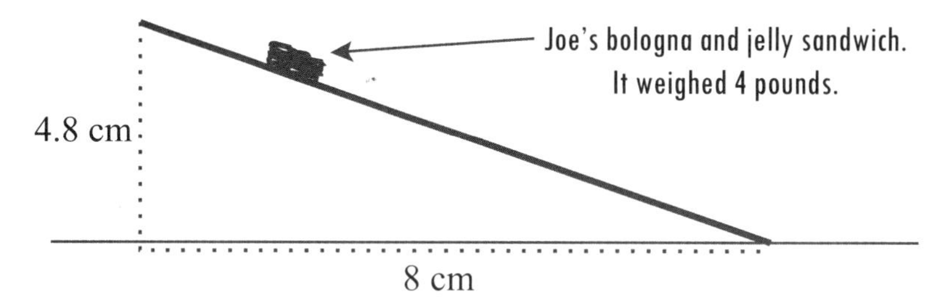

Find the coefficient of static friction.

9. When Joe was chewing on his bologna and jelly sandwich, he noticed something funny. There was a rubber band in his jelly. He couldn't figure out how it got there. (Things like that happened to Joe all the time. His life was marked by chaos.)

He played with the rubber band. As he stretched it, he noticed that as long as he stayed within the proportional limits, the amount of stretch (inches) was proportional to the force (pounds) he used. What is the name of that law?

10. If $5x = 2$, what does x equal?

The Bridge

from Chapters 1–37

second try

1. Darlene thought that her wedding cake should be very special. At first, she thought of putting candles on the cake but decided that would look too much like a birthday cake.

Instead, she would have a row of electric lights strung across the bottom of the cake.

The three lights would each be 5 ohms. If they were in series, what would be the total resistance?

2. If the lights were powered by a 12-volt battery, what would be the current flowing between the first and second lights?

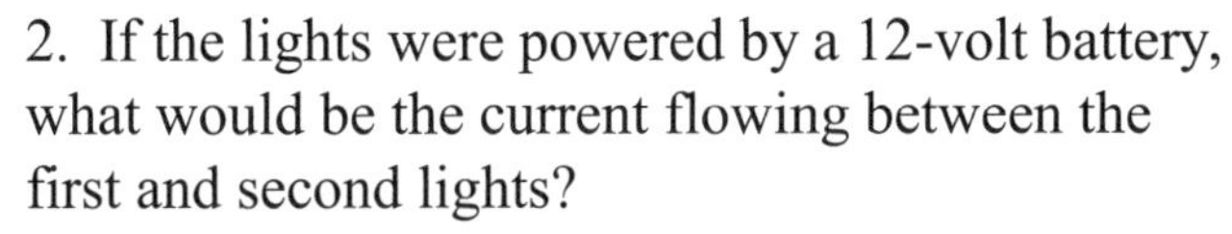

3. If the three lights were in parallel, what would be the total resistance?

4. If the cake were made with 50 pounds of sugar, 20 pounds of flour, and 10 pounds of metal decorations, what percentage of the cake would be sugar?

5. There would be a swimming pool at the wedding reception site so that the guests could play in the water if they wished before the wedding. So that they wouldn't forget why they were there, Darlene would have a copy of the wedding cake made and placed at the bottom of the pool. Since the original cake would dissolve in the water, she would have a copy made out of steel.

The 620-pound steel cake would be delivered to the edge of the pool. The coefficient of static friction between the cake and the concrete at the edge of the pool was 0.7. How much force would be needed to get the cake moving?

6. If the guests dove down to the submerged cake and tried to lift it, how much would it seem to weigh?

Volume of the steel cake = 2 ft^3. Weight = 620 lbs. Density of the pool water = 62 lbs./ft^3.

7. Darlene read in a bridal magazine (and believed) that 70% of people who are invited to a wedding actually show up. She planned on sending

out a thousand invitations. She figured that 700 people would attend her wedding. Is this an example of deductive or inductive reasoning?

8. Darlene imagined that the best way for people to remember whose wedding it was would be for the baker to put her name in pink frosting all over the wedding cake.

The baker told her that the standard rate for lettering is $3 for every 12 letters on the cake. Darlene wanted 16 copies of her name on the cake. How much would that cost? (This can be done with a conversion factor.)

9. Is the number of letters that the baker puts on the cake a discrete or a continuous variable?

10. Which is the larger numeral: 8 or 5 ?

The Bridge

from Chapters 1–37

third try

---

1. Christina Rossetti was one of Fred's favorite poets. One evening he was reading "Advent" and found the line:

We weep because the night is long. . . .

Fred looked at the three lightbulbs that were on in his office. Rossetti was born in 1830. That was before lightbulbs were popular.*

If Fred's three lightbulbs, which are in parallel, had resistances of 20 Ω, 30 Ω, and 40 Ω, what would be the combined resistance of all three?

2. Fred turned on his computer. According to the ammeter he had attached, 2 amps ran through the computer. If the voltage was 110 volts, what was the resistance of his computer?

3. Fred found a stack of bologna slices in his desk drawer. Each slice was very thin and had an area of 4 in^2. The stack was 7-inches tall. What was the volume of that stack?

4. There are two usual ways to measure the mass of an object. One way is to put it on a scale and compare it with the weight of other objects whose mass you know.

What is the other way to measure mass?

5. Fred could read Rossetti's poems at the rate of 6 poems every 8 minutes. At that rate, how long would it take him to read 45 of her poems? (Use a conversion factor.)

6. The slices of bologna had been in Fred's desk drawer for years. They were completely dried out. He took one slice (0.07 pounds) and placed it on his desk. It took 0.04 pounds of force to move it at a constant speed across his desktop. Find μ_k.

7. Given the information in the previous problem, is it possible to find μ_s?

---

* Thomas Edison introduced his lightbulb to the public in 1879. He said, "We will make electricity so cheap that only the rich will burn candles."

The Bridge

from Chapters 1–37

8. Fred read the last stanza of "In the Willow Shade":

> ***That silvery weeping willow tree***
> ***With all leaves shivering,***
> ***Which spent one long day overshadowing me***
> ***Beside a spring in Spring.***

He knew from the first lines of the poem—"I sat beneath a willow tree/Where water falls and calls"—that "spring" meant creek rather than some stretchy piece of metal.

He giggled and mentally pictured

You know how five-year-olds are. If he pulled on the spring with a force of 2.4 pounds, it would stretch 6 inches. How much force would it take to stretch it 10 inches?

9. If he pulled the spring 20 inches, it would get permanently bent out of shape. Which of these four applies to that situation? A) The spring would have reached its proportional limit; B) It would have reached its elastic limit; C) It would have entered the plastic region; or D) It would have passed the breaking point.

10. Every time Fred giggled he noticed that he felt better. He concluded that giggling made him feel better. Is this an example of inductive or deductive reasoning?

The Bridge
from Chapters 1–37

fourth try

---

1. It was Jake, Jack, and Jane's fifth day in New York City. It was starting to snow so they decided to do something indoors. Jane suggested, "Let's visit a museum. New York City has a lot of them."*

"Can we hit all of them today?" Jake asked.

Jane giggled. "Let's just go to one museum, the Met. We won't be able to see everything at the Met even if we spend all day there."

Jake was confused. "The Met—isn't that a baseball team?"

"No," said Jane. "The Met is what everyone calls the Metropolitan Museum of Art."

Jack was dragging his bag full of electronic stuff that he had bought during the first four days in New York City. The bag weighed 90 pounds. The coefficient of kinetic friction between the bag and the sidewalk was 0.7. How hard did Jack have to pull to keep his bag moving at a constant speed?

2. Jack didn't want to drag his bag through the museum. He found a locker and lifted his bag 3 feet upward to put it inside. How much work did that involve?

3. Jack could move a lot faster now that he wasn't dragging his purchases. The three sailors raced to the entrance to the Met on Fifth Avenue. Jack could not believe his eyes. The Met is one of the largest museums in the

---

* Jane was not kidding. Here is a list of some of them. Alice Austen House Museum, American Craft Museum, American Folk Art Museum, American Museum of Natural History, American Museum of the Moving Image, American Numismatic Society, Americas Society, Artists Space, Asia Society and Museum, Bronx Museum of the Arts, The Brooklyn Botanic Gardens, The Brooklyn Children's Museum, The Brooklyn Museum of Art, Carnegie Hall/Rose Museum, Central Park Zoo/Wildlife Gallery, The Children's Museum of the Arts, Children's Museum of Manhattan, The Cloisters, Cooper-Hewitt, Dahesh Museum, Dia Center for the Arts, The Drawing Center, Ellis Island Museum, Empire State Building Lobby Gallery, Museum at FIT, Forbes Magazine Galleries, The Frick Collection, Grey Art Gallery, Goethe House, Solomon R. Guggenheim Museum, Guggenheim Museum SoHo, Rose Center, The Hispanic Society of America, International Center of Photography, Intrepid Sea-Air-Space Museum, Isamu Noguchi Garden Museum, Jewish Museum, LaGuardia and Wagner Archives, Lower East Side Tenement Museum, Madame Tussaud's New York, Merchant's House Museum, Metropolitan Museum of Art, The Morgan Library, Mount Vernon Hotel Museum & Garden, Municipal Art Society, El Museo Del Barrio, Museum at Eldridge Street, Museum for African Art, Museum of American Financial History, Museum of Chinese in the Americas, Museum of Jewish Heritage, Museum of Modern Art, Museum of the City of New York, Museum of the Moving Image, Museum of Television and Radio, National Academy Museum, National Design Museum, National Museum of the American Indian, New Jersey Children's Museum, New Museum of Contemporary Art, New York Botanical Garden, New York City Fire Museum, New York City Police Museum, New York Hall of Science, New York Historical Society, New York Public Library, New York Transit Museum, Nicholas Roerich Museum, PS1 Contemporary Art Center, Pierpont Morgan Library, Queens Historical Society, Queens Museum of Art, Rose Center for Earth and Space, Schomburg Center, Snug Harbor Cultural Center, South Street Seaport Museum, Staten Island Institute, Studio Museum in Harlem, Taipei Gallery, Theodore Roosevelt Birthplace, Ukrainian Museum, Wave Hill, Whitney Museum of American Art.

world. The entrance has water fountains, huge stairways, and tall columns.

Jack estimated that one of the columns was 40-feet high and the base of the column had an area of 2 square feet. What was the volume of that column?

4. They each received a map as they entered the Met. Jake gasped as he looked at the map and said, "This joint is huge."

Each gallery was one or more rooms. There were 60 galleries devoted to European paintings from 1230 to 1900.* Those 60 galleries are what percentage of all the 400 galleries at the Met?**

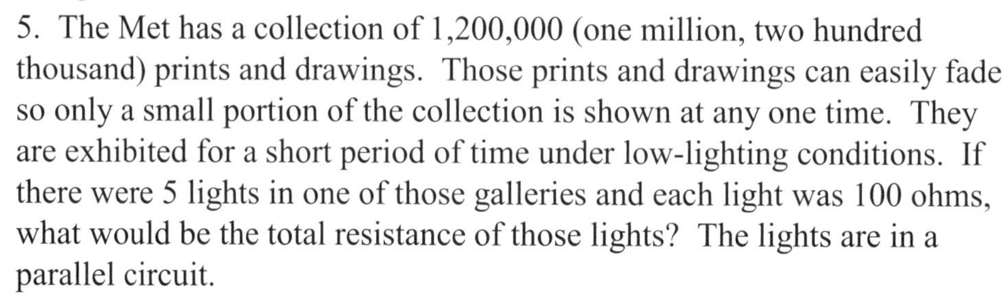

5. The Met has a collection of 1,200,000 (one million, two hundred thousand) prints and drawings. Those prints and drawings can easily fade so only a small portion of the collection is shown at any one time. They are exhibited for a short period of time under low-lighting conditions. If there were 5 lights in one of those galleries and each light was 100 ohms, what would be the total resistance of those lights? The lights are in a parallel circuit.

6. The Egyptian Art gallery has a whole building inside of it. It is an Egyptian temple that was shipped to America as a gift. If the temple weighed 10 tons and it cost $55,830 to ship each ton, using a conversion factor, determine the cost of shipping the whole temple. (No credit is given for this problem if you don't use a conversion factor.)

7. Is the number of galleries a discrete or continuous variable?

8. If $6 = 18y$, what does y equal?

9. $\sqrt{49} = ?$

10. Jake had thrown a penny into the water fountain outside the Met. The penny sank to the bottom (4 feet). What is the pressure on that penny? Call the density of water 62 pounds per cubic foot.

---

* The modern art, from 1900 to the present, is housed in 30 more galleries.

** Actually, there are a few more than 400 galleries, but I want to keep the arithmetic simple.

The Bridge

from Chapters 1–37

fifth try

1. Stanthony invented the Six-Topping Pizza™. (Someone else had already invented the One-Topping Pizza™, the Two-Topping Pizza™, . . . , the Five-Topping Pizza™, and the Seven-Topping Pizza™.) Is the number of toppings on a pizza a continuous variable?

2. Sixty bell peppers cost $9. Using a conversion factor find out how much 400 bell peppers cost.

3. After the Six-Topping Pizza™ is baked, Stanthony keeps it warm under three heat lamps. The lamps are in parallel and have resistances of 6 Ω, 8 Ω, and 12 Ω. What is the total resistance of those three lamps?

4. If those three lamps were in series rather than parallel, what would be their total resistance?

5. Given this diagram, find I.

12 Ω

6 V

\+ −

A

6. If 8x = 3, find x.

7. Which is the larger number:
   4 or 5?

8. The Six-Topping Pizza™ is heavy. The toppings are piled so high that it exerts a pressure of 2 pounds per square inch on the table. What is the pressure in terms of pounds per square foot?

9. Stanthony's Big Party Table™ is rectangular. It measures 5 feet by 26 feet. What is its area?

10. Stanthony put a medium-sized Six-Topping Pizza™ on his Big Party Table™. That pizza is designed to serve 20 hungry football players. It weighs 80 pounds.

When one player at one end of the table yelled, "Hey! Pass the pizza!" another player at the other end simply lifted his end of the table up until the pizza began to slide. Find μ_s.

(Lifting a table up ten feet is not hard for a football player who is seven-feet tall.)

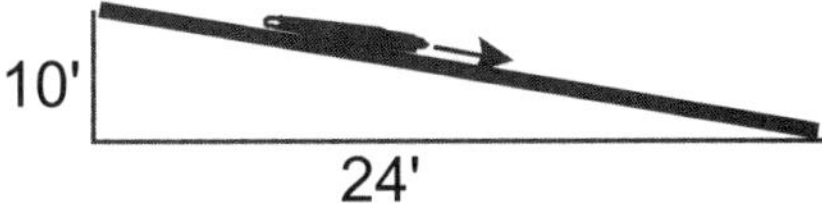

The Bridge

from Chapters 1–37

sixth try!!!!!

Wait! I, your reader, need to call a time out. You have never had six tries on a Bridge. What's this all about?

I'm in the mood to do something crazy.

I don't like the sound of this.

All week I have been thinking about electrical circuits. In the chapters we have just covered, we took some baby steps. For example, we looked at

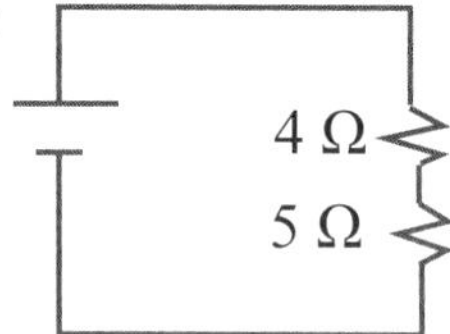

mentally circled a section,

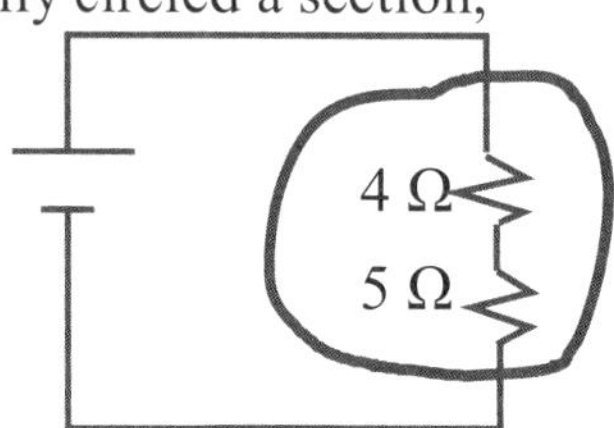

and replaced it by

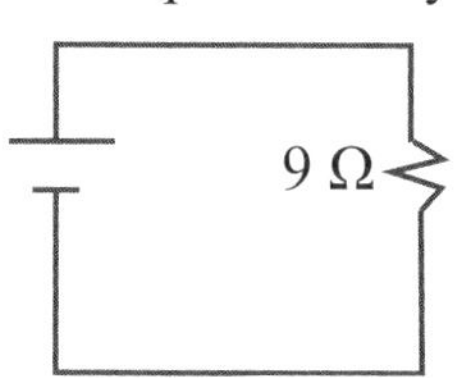

We also did it with parallel circuits. In the solution to #2 on page 214, we mentally circled

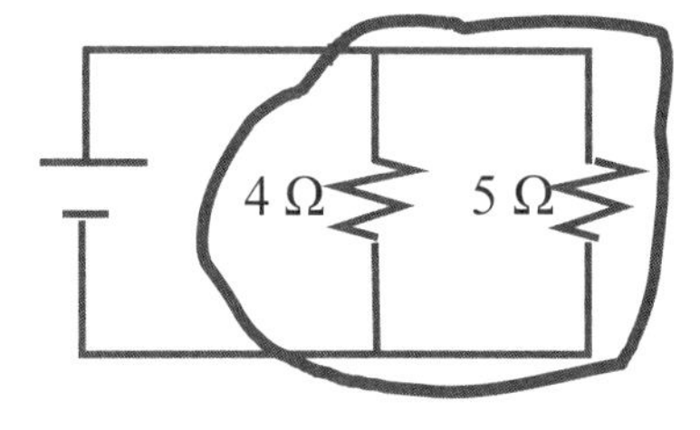

and replaced it with

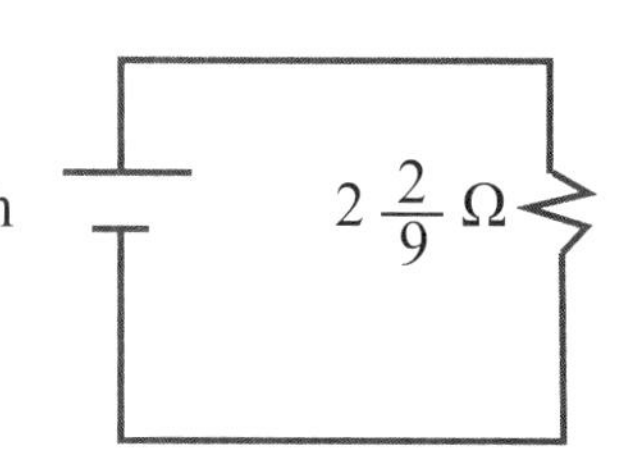

The trick with circling is to have exactly one line going into the circle and one line coming out.

The Bridge

from Chapters 1–37

You are leading up to something. I can tell.

Actually, I'm leading up to four different things.

First: This is the last try of the last Bridge in this book.

What about a Final Bridge?

These six tries are the Final Bridge.

But what about the rest of the book? You have about twenty pages to go before we hit the index.

Second: What lies ahead is pure eye candy. There will be no new laws of physics, no computation, and no conversion factors—just a story. It will be the dessert at the end of this book. You will learn things that most adults don't know.

Third: This sixth try is different than any other Bridge you have done. I want to do something outlandish: a Bridge try with a single question. Get it right and you pass the Bridge.

I bet the question is insanely difficult.

This Bridge try is my gift to the 0.382% of my readers who are super-gifted and are thinking of becoming EE majors.* The rest of my readers (99.618%) can just look at the question on the next page and laugh.

Two kinds of books are bad: books in which you can understand nothing and books in which everything is too easy. This sixth try will eliminate the second alternative.

Fourth: The answer to the question on the next page is 6 amps.

Wait! A one-question Bridge and you are giving the answer?! This is truly frightening. I refuse to turn the page.

It is your choice.

---

* That is what electrical engineering majors are called in college.

The Bridge
from Chapters 1–37

You turned the page.

This is the sixth try of the last Bridge in this book. Here is the question: Find the current, I, through the ammeter.

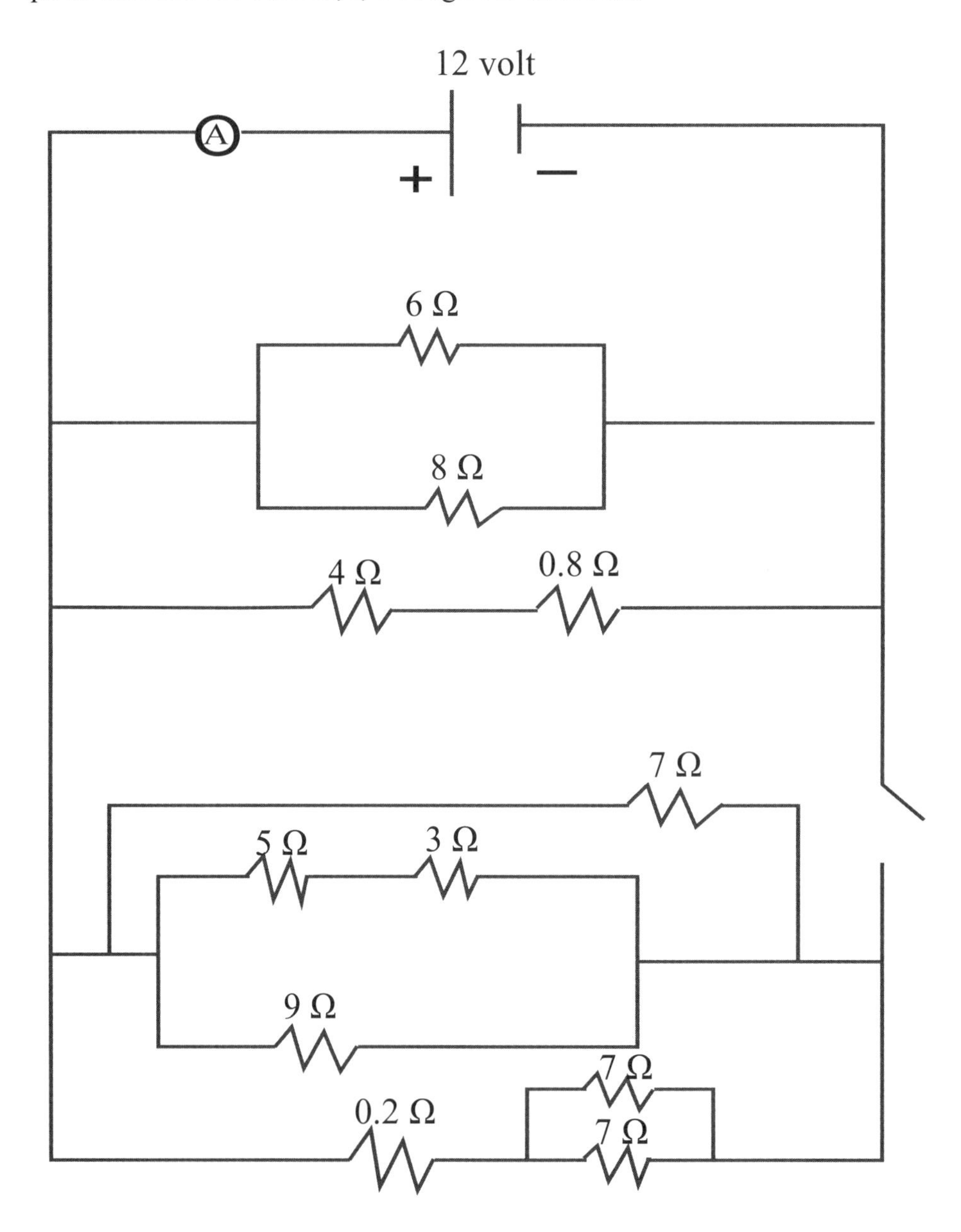

Chapter Thirty-eight
Lunch

Kitty liked playing with all the electric gadgets in the kitchen. She popped a couple of English muffins into the toaster, put eight slices of bacon into the microwave, and made a giant chocolate shake in the blender.

"It's time we break for lunch," she announced.

Lunch Fred thought to himself. The only thing I have done as chief electrician is put some batteries in a flashlight. She already has had a huge breakfast: hamburger, onion rings, and a strawberry milkshake.

Fred looked at the clock. It was ten o'clock.

He headed to the refrigerator and got some butter and some grape jam for her English muffins. When he got back, he found that she had set the table for both of them. Kitty had put four slices of bacon and two halves of an English muffin on each of their plates. She had made a quart of chocolate milkshake. Each of them had a pint.

They sat down and Kitty dove into her food.

"Eat up!" she told him. "No need to be bashful. I know you are a growing boy."

Fred looked at the mountain of food in front of him. He knew he had already had breakfast here at the café.* Now in front of him was more food than he might eat in a year. There are two kinds of prayers that God especially enjoys hearing: prayers of thankfulness and prayers asking for guidance. Fred quietly said both. He was grateful that he had a job as Chief Electrician, and he really needed to know what to do in this situation.

He did what every five-year-old does when he is not really hungry. He played with his food. He cracked one of the pieces of bacon in half to make eyes and used the other three pieces to make the mouth. The muffins became ears. He called it Muffin Mouse.

---

* This was Fred's definition of breakfast: an attempt to suck some lake water up a straw.

He took an eye off of Muffin Mouse. That was too much bacon to eat. He broke it in half and put part of it back on his plate.

Still too much*. He continued breaking it in half. Four grams became 2 grams, became 1 gram, became 0.5 grams, became 0.25 grams, became 0.125 grams, became 0.0625 grams, became 0.03125 grams, became 0.015625 grams, became 0.0078125 grams, became 0.00390625 grams, became 0.001953125 grams, became 0.0009765625 grams, became 0.00048828125 grams, became 0.000244140625 grams, became 0.0001220703125 grams, became 0.00006103515625 grams, became 0.000030517578125 grams, became 0.000015258789063 grams, became 0.000007629394531 grams.

A fly came, grabbed that piece of bacon, and flew away with it.

small essay

How Many Atoms Are In That Tiny Piece of Bacon?

Bacon is composed mostly of carbon (atomic weight = 12), oxygen (atomic weight = 16), and hydrogen (atomic weight = 1). You can find atomic weights in a periodic table of the elements in any chemistry book.

Chemistry facts:

6×10^{23} atoms of carbon weigh 12 grams.

6×10^{23} atoms of oxygen weigh 16 grams.

6×10^{23} atoms of hydrogen weigh 1 gram.

This is true for any element: 6×10^{23} atoms of an element is equal to its atomic weight.

One atom of oxygen weighs $\frac{16 \text{ grams}}{600{,}000{,}000{,}000{,}000{,}000{,}000{,}000}$

If each atom in the bacon were oxygen (the heaviest of the three—carbon, oxygen, and hydrogen), using a conversion factor:

$$\frac{0.000007629394531 \text{ grams}}{1} \times \frac{600{,}000{,}000{,}000{,}000{,}000{,}000{,}000 \text{ atoms}}{16 \text{ grams}}$$

* This is an elliptical construction. Filling in all the words: *This was still too much.*

equals 286,102,294,912,500,000 atoms in that tiny bit of bacon.

Two hundred eighty-six quadrillion, one hundred two trillion, two hundred ninety-four billion, nine hundred twelve million, five hundred thousand atoms in that 0.000007629394531 grams of bacon.

Atoms are small.

end of small essay

Fred knew that it wasn't polite to talk with food in your mouth, so he figured that if he were speaking, he would not be expected to be eating.

Fred knocked on the table and asked, "Kitty, do you know what this is made of?"

Kitty took a swallow of her chocolate shake and said, "Wood."

"And what is wood made of?" Fred continued.

"You tell me." Kitty was more interested in eating than talking. This was Fred's opportunity. This is his . . .

History of the Atom

430 BC

Once upon a time, there lived a Greek whose name was Democritus. (deh-MOCK-ri-tuss) He asked that same question: What is everything made of?

He just sat and thought about what things were made of. He didn't look at nature. He guessed that everything was made of atoms that are:

- ☆ small
- ☆ hard
- ☆ indestructible, and
- ☆ infinite in number.

He was right about them being small. He was wrong about everything else. He was 25% right.

1704

Two thousand four hundred years pass. Isaac Newton wrote, ". . . it seems probable to me that God in the Beginning form'd Matter in solid, massy, hard, impenetrable, moveable Particles. . . ."*

Not much progress had been made.

1896

Henri Becquerel studied the radioactivity of the uranium atom. At this point, scientists were actually starting to really look at nature. The uranium atom was not stable. It fell apart. Goodbye Democritus. Goodbye Newton. The atom was not indestructible.

1897

A year later, by experimenting, J. J. Thomson discovered the electron. It was part of an atom. It had a negative charge. Since atoms were electrically neutral, the positive and negative charges must balance. Atoms were not hard, featureless balls. They had a structure.

In 1904 he imagined that atoms were clouds of positive charge with bits of negative charge floating in those clouds like—as he suggested—raisins scattered in a plum pudding. Nowadays, it is more popular to think of blueberries in a blueberry muffin or chocolate chips in a chocolate chip cookie.

1911

Ernest Rutherford blew Thomson's plum pudding all to bits in a famous experiment. He put some polonium in a lead box with a hole in it. Polonium is radioactive, and when it decays, it spits out alpha particles.

He aimed this "gun" at some super-thin gold foil.

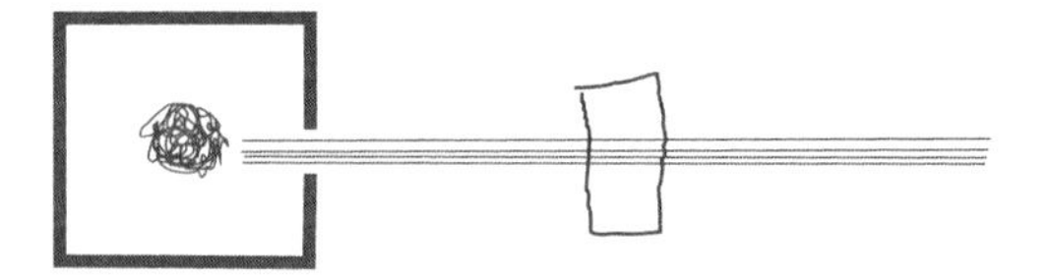

* Three centuries ago, spelling and capitalization were a little different than now. Newton wrote *form'd* rather than *formed*.

Alpha particles are about 7,000 times more massive than electrons. If Thomson's plum pudding model was true, then those alpha particles would pass through those thousands of gold atoms and be deflected by one or two degrees at most.

Short geometry lesson:

90° 45° 5° 1°

Imagine firing a rifle into a glob of pudding. Or better yet, into a piece of tissue paper. This is the situation that Thomson's plum pudding model was predicting.

Rutherford asked nature, "Are you plum pudding?" He got the surprise of his life. Some of the alpha particles bounced right back at him. Goodbye to Thomson's plum pudding.

His experiment showed that atoms had to be mostly empty space with a tiny positive nucleus. The nucleus was one-trillionth of the total volume of the atom.*

Rutherford pictured a super-small nucleus with electrons flying around it in circular orbits. If the period at the end of this sentence were the nucleus, then the electrons would be in another room of your house.

Fred drew on a napkin (while Kitty kept eating):

Mr. Democritus	Isaac Newton	Henri Becquerel	J. J. Thomson	Ernest Rutherford
small, hard, indestructible, infinite	solid, massy, hard	they can fall apart	discovered the electron, cloud pudding	small nucleus, planetary model, circular orbits
430 BC	1704	1896	1897	1911

It would have been better if Kingie had done the drawings.

---

* Translation: Like a fly in a cathedral or a marble on a football field.

Kitty stopped eating for a moment and looked at Fred's napkin. "I don't know much about atoms, but Democritus and Fig Newton looked like the only ones who knew what was going on."

Isaac Newton, not Fig Newton, Fred thought to himself.

Kitty continued, "Do you remember that golf ball I accidentally whacked you with?"

Fred rubbed his head.

She asked him, "Did the atoms in the golf ball fall apart? Did getting hit by the golf ball feel like being hit with plum pudding? Does a golf ball feel like it is made of atoms that are almost completely empty space?"

Fred shook his head.

"Those guys—fall-apart Becquerel, pudding Thomson, and circular-orbit-empty-space Rutherford—do not know anything about golf balls."

Fred waited until Kitty put a slab of butter and a small handful of grape jam on an English muffin and took a big bite. He could then continue his history of the atom.

He drew a time line:

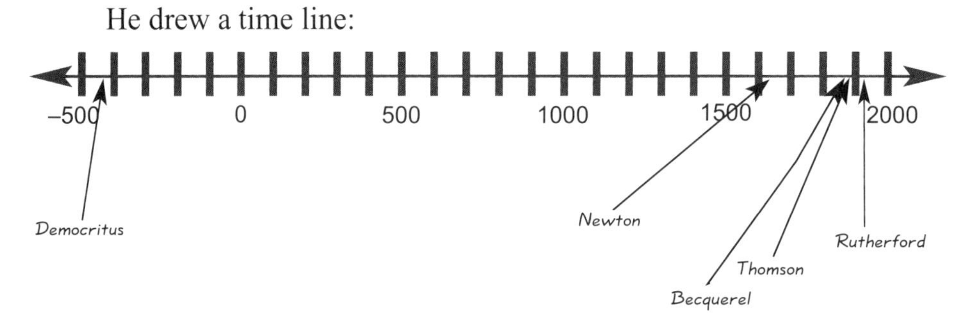

In the space of 15 years (1896–1911), our idea of the atom changed a lot.

In the next 21 years (1911–1932), we would move from the merely strange—atoms are mostly empty space—to THE UNBELIEVABLE.

Chapter Thirty-nine
From Strange to Stranger

Fred was afraid to continue the history. He was worried that Kitty would think that he was just making up a story rather than telling what really happened. When Ernest Rutherford showed with his polonium "gun" and gold foil experiment that the nucleus was one-trillionth ($\frac{1}{1,000,000,000,000}$) of the volume of an atom, many people refused to believe it.

But there was something cute about Rutherford and his model of the atom.

1913

Just two years after Rutherford ate Thomson's plum pudding, Niels Bohr put an end to Rutherford's happy dance. No longer could electrons pick *any* distance away from the nucleus. Bohr said they had to march in precise orbits, at predetermined fixed distances. The closest that an electron could get was n = 1. The next closest orbit he called n = 2. The next slot was n = 3. The next track was n = 4. The next energy state was n = 5. The next shell was n = 6. The next principal quantum number was n= 7.*

You will never find an electron flying around the nucleus at a distance of n = 4½ or n = 5.60923. They are locked into fixed tracks.

---

* English: slot = track = energy state = shell = principal quantum number.
Math: n = 1, 2, 3, 4, 5, 6, 7, 8, 9. . . .
You tell me which is easier: English or math.

In the language of math, n is a discrete variable, not a continuous one.

If Thomson was pudding and Rutherford was a free dance at great distances, then Bohr was Click! Click! The electrons had to march in formation like a band on a football field or soldiers on parade.

If an electron is orbiting in energy state n = 4 and it decides it wants to move to lower energy state n = 3, it does not slide like this:

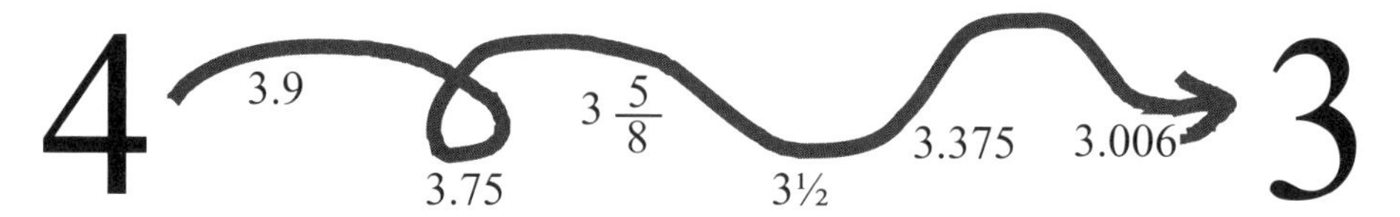

Instead, it clicks. One moment it has a principal quantum number equal to 4. The next moment it is in shell number 3.*

When it clicks down to a lower energy state, it will give off a bit of light. Fred said, "A bit of light." He could have said, "It emits a quantum packet of electromagnetic radiation," but he wanted to keep his language simple so that it didn't disturb Kitty's enjoyment of her lunch.

Actually, nothing could disturb her eating. Kitty was now eating the bacon off of Fred's plate.

Wait a minute! Stop! I, your reader, have to know something. How did Click! Click! Bohr—as you call him—ever come to the conclusion that electrons can only be in certain orbits. Was he just making up what he thought was true like Democritus invented his "facts" about atoms?

If there are 286,102,294,912,500,000 atoms in a 0.000007629394531-gram crumb of bacon, you can bet your boots that Bohr didn't see those orbits.

No he didn't. But he did see the light.

---

* Fred, at the age of five, is single. Some day he will probably be married. His quantum state will change, but at no point will he be 30% single and 70% married.

What light? Or are you speaking metaphorically?*

Recall that Fred said that when an electron drops down to a lower energy level, it gives off a bit of light. Bohr saw that light.

How? When? I don't see any light.

Hydrogen is the simplest atom: one electron circling one proton in the nucleus. When an electron in a hydrogen atom jumps from the orbit n = 4 to the orbit n = 2, it emits a blue light—not just any shade of blue like azure mist, Alice blue, baby blue, periwinkle, powder blue, Cornflower blue, sky blue, aquamarine blue, turquoise blue, Ukrainian Azure, United Nations azure, cerulean, Bondi blue, steel blue, agate blue, indigo, slate blue, Dodger blue, royal blue, denim, Swedish azure, cobalt blue, Persian blue, lavender, International Klein blue, Ultramarine, navy blue, sapphire, midnight blue, Prussian blue, teal, Palatinate, Federal blue, Phthalo blue, or Air Force blue—but one particular shade.

Here is what Bohr did to earn the name Click! Click! Bohr that I have given him.

He filled a glass tube with a gas and passed an electric current through it. It took a high voltage to do that since gas does not conduct electricity as easily as a copper wire.

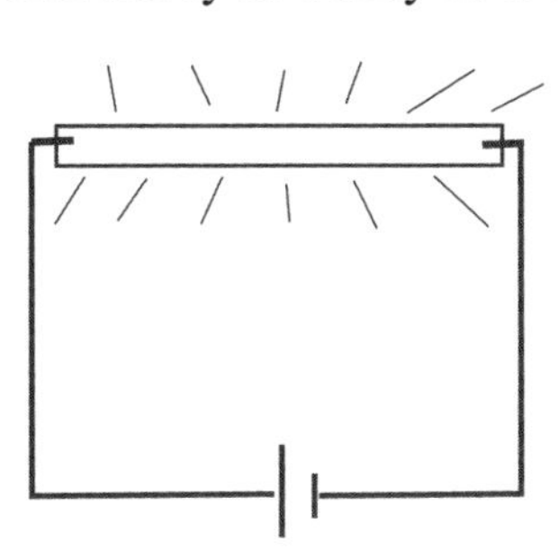

The electricity got the gas atoms all excited. The electrons hopped between the different orbits.

If Rutherford was right and orbits can be at *any* distance from the nucleus, then the tube would have shown every color of light—every shade of blue and red and yellow and green and. . . .

---

* "Speaking metaphorically" means to use a metaphor.

I have to first define *simile* (SIM-a-lee). A simile compares two things. For example: Kitty was eating like a hog.

A metaphor declares that something *is* something else. For example, Kitty was being a hog.

When I wrote, *Ernest Rutherford blew Thomson's plum pudding all to bits*, that was a metaphor.

Marching electrons is a metaphor. When we learned from Bohr that the orbits of electrons were *like soldiers on parade* we are using a simile.

Metaphors are more compact than similes. They have more punch. [⇠ a metaphor] Metaphors add spice to your writing. [⇠ another metaphor]

But that didn't happen. When Bohr excited hydrogen, it gave off very distinct colors: 656.6 nm, 486.1 nm, 434.1 nm, 410.2 nm, etc. These colors corresponded to electrons clicking between the shells n = 1, n = 2, n = 3, n = 4, etc.

Hold it! I have a couple of questions. Actually, three questions.* ***First of all, what does "nm" mean?***

It is an abbreviation from the metric system. In that system, m stands for meter, cm stands for centimeter (one-hundredth of a meter), and nm stands for nanometer (one-billionth of a meter).

You said that hydrogen gives off the color 486.1 nm. This seems crazy. How can a color be a length?

This is something you will never learn in an art class. Light comes in waves like a ripple on a pond when you throw in a golf ball. The color of a bit of light depends on how far apart the waves are from each other.

At one end of the rainbow (the red end), the waves are far apart, and at the other end of the rainbow (the violet end), the waves are closer to each other.

The shades of blue are from 455 nm to 492 nm. The shade of blue that hydrogen gives off (wavelength = 486.1 nm) is only one of the hundreds of possible shades of blue.

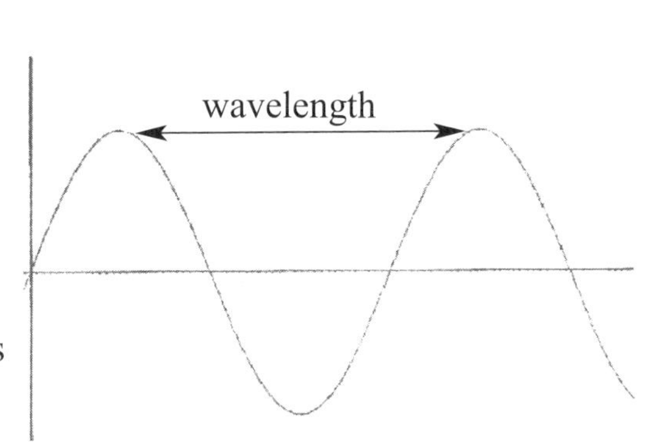

---

* In everyday life, people use elliptical constructions like this all the time when they are speaking. When you are writing, you should avoid using elliptical constructions. *Actually, three questions* is called a sentence fragment. One English teacher I had in college would flunk any essay that had a sentence fragment in it. If this were a formal essay, I would have written *Actually, I have three questions.*

I learned very quickly never to write a sentence fragment. Hardly ever.

So if you had a box of crayons and you were a physicist, you could label them by their wavelengths instead of their color names.

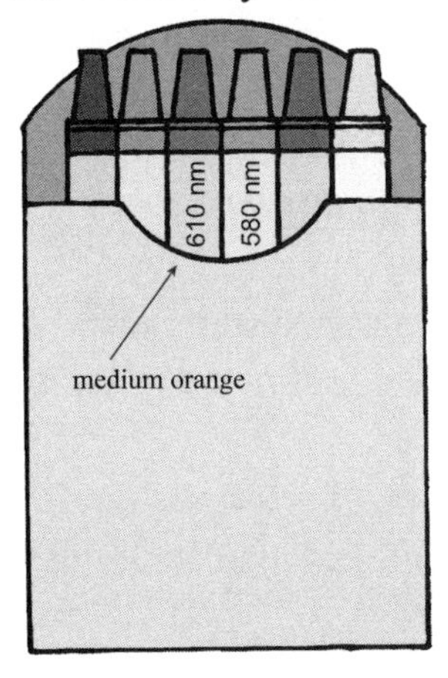

Handy Chart

Violets come in shades between 390 nm and 455 nm.
Blues come in shades between 455 nm and 492 nm.
Greens come in shades between 492 nm and 577 nm.
Yellows come in shades between 577 nm and 597 nm.
Oranges come in shades between 597 nm and 622 nm.
Reds come in shades between 622 nm and 770 nm.

My third question is how in the world did Bohr ever determine that the blue light coming from an electron jumping from the n =4 shell to the n = 2 shell in a hydrogen atom was 486.1 nm? Did he hire an artist like Kingie to say it was 486.1 nm and not 486.2 nm?

Even an artist like Kingie couldn't do that.

"It looks like a medium blue."

The easy way to measure the wavelength is to use a device called a spectrometer (speck-TROM-eh-ter). You can buy one on the Internet. I saw one for about $1,800 that will measure wavelengths from 190 nm to 1100 nm and has a mass of 44 kg (which weighs about 97 pounds).

It is fairly certain that Niels Bohr didn't buy his spectrometer on the Internet in 1913.

Wait a minute. That spectrometer that you can buy on the Internet can measure wavelengths up to 1100 nm. But what color is 1100 nm? It is far beyond the reds, which go from 622 nm to 770 nm.

It isn't a color. At least it isn't any color that we can see. Our eyeballs can only work in the range from about 390 nm to 770 nm.

Just because we can't see something doesn't mean that it doesn't exist. (← Important point)

Do you remember four pages ago when Fred did not mention electromagnetic radiation to Kitty? Instead, he called it a bit of light. Electromagnetic (six syllables!) waves are not limited to 390 billionths of a meter to 770 billionths of a meter. (0.00000039 to 0.00000077 meters)

If you have ever broken a bone and the doctor X-rayed it, the X-rays may have had a wavelength of about 1 nm (0.000000001 meter).

On the other hand, how about a wavelength of 60 meters? (A meter is a little longer than a yard.) Those nice big long waves are what your radio receives. I'm not sure how such a long wave fits inside of your little radio—but it does.

Any more questions, or can we let Fred continue his history of the atom?

I'm cool—for now. Hold it. I just thought of one.

Are neon lights. . . .

Yes. Exactly. Electricity through glass tubes filled with gas.

So Fred continued by telling Kitty that the atoms of each element are different sizes so when an electron makes a jump from the n = 4 shell to the n = 2 shell in a hydrogen atom it gives off a different wavelength than the same jump in an oxygen atom.

If the spectrometer says 656.6 nm, 486.1 nm, 434.1 nm, 410.2 nm you know you are looking at a hydrogen atom.

Before Bohr's time, they pointed a spectrometer at the sun and found lots of the same elements that are here on earth. In 1868, Lockyer found the wavelengths of something that didn't exist on earth. In Greek, the sun is called *helios.* They named this new substance *helium.*

Later, scientists found helium in certain natural gases in the United States. The very first place was in Kansas—the home state of KITTENS University where I teach.

During World War I (1914–1918), helium cost about $2,000 per cubic foot. In 1955 the price had dropped a little. It was 9¢ per cubic foot.

Nowadays, every element that has been found in starlight has also been found on earth.

Chapter Forty

Stranger to Unbelievable

Kitty asked Fred if he was willing to share one of his English muffin slices. Fred nodded and warned her that his history of the atom was going to get even more weird.

"What could be more weird than what you've told me so far?" Kitty asked. "You have Rutherford saying that atoms are mostly empty with the little electron balls a zillion miles away from a super-small nucleus and they are flying around in circular orbits. Then in 1913 Click! Click! Bohr determines that those electrons can only fly in particular orbits, which he names n = 1, 2, 3, 4, 5, 6. . . ."

Fred continued his story.

a few months after Bohr's 1913 announcement

Bohr's discovery of the energy states of an electron did not last long. Arnold Sommerfeld liked Bohr's shells (n = 1, 2, 3, 4, etc.), but he noticed that an electron in the n = 2 shell could have 2 possible orbits. Orbits could be elliptical (in the shape of an ellipse) rather than circular.

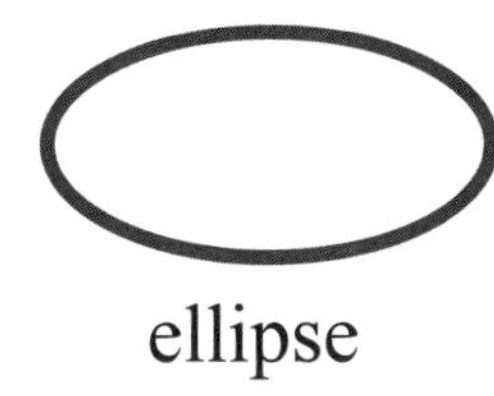

ellipse

The two subshells for the n = 2 shell, he named ℓ = 0 and ℓ = 1. We use a cursive ℓ rather than a regular l. One (1) and an l look too much alike.

Sommerfeld found that there were three subshells (ℓ = 0, 1, and 2) for the n = 3 shell.

There were four subshells (ℓ = 0, 1, 2, and 3) for the n = 4 energy level.

In short, the nth shell has n subshells.

n = 1 shell has 1 subshell. ℓ = 0
n = 2 shell has 2 subshells. ℓ = 0, 1
n = 3 shell has 3 subshells. ℓ = 0, 1, 2
n = 4 shell has 4 subshells. ℓ = 0, 1, 2, 3

"What a mess!" Kitty exclaimed.

Fred continued.

1923

Louis de Broglie was thinking about the fact that light sometimes acts like a wave (electromagnetic radiation) and sometimes like a particle, which physicists call a photon.

He said that if light can play both games (wave/particle), then it seems only fair that matter can also sometimes act like a particle and sometimes like a wave.

Kitty had just finished Fred's second slice of English muffin. She said, "This is getting silly. My golf ball can be a wave? Physicists believe stuff that is really . . . unbelievable."

Fred wanted to mention that in 1927 C. J. Divisson and L. H. Germer measured the wavelength of an electron. He didn't because that would be getting ahead of the story. He continued with:

1925

Let's get back to the orbits (shells, energy states) of an electron. Bohr had invented the principal quantum number n to describe which shell an electron could be in.

Sommerfeld invented the **orbital quantum number ℓ** to describe which subshell an electron could be in. At this point, a particular electron could live, for example, at $n = 7$ and $\ell = 5$.

Then physicists found that there were sub-subshells! They invented the **orbital magnetic quantum number** m_ℓ to number the various possible sub-subshells.

Then two graduate students at the University of Leiden in The Netherlands got a really nice spectrometer. It was one that could really measure wavelengths accurately. Samuel Goudsmit and George Uhlenbeck made one more subdivision of the electrons in orbits around the nucleus. They were running out of words to describe things and settled on **spin magnetic quantum number** that they called m_s.

At this point, n, ℓ, m_ℓ, and m_s described the various energy levels of an electron in an atom.

Kitty said, "I think I'm going to throw up." She was imagining what would come after n, ℓ, m_ℓ, and m_s.

later in 1925

Wolfgang Pauli said, "Enough!" He created the Pauli exclusion principle: No two electrons in the same atom can have the same four numbers: n, ℓ, m_ℓ, and m_s. It is the same as no two people can have the same Social Security number.*

Roughly a century has passed since Pauli created his exclusion principle. Many other ideas about the atom have turned out to be incorrect. Democritus was wrong. Newton was wrong. Pudding Thomson was wrong. Circular orbit Rutherford was wrong. Click! Click! Bohr wasn't right when he thought that n = 1, 2, 3, etc. would completely describe the orbits of a electron.

Pauli appears to have been right and strangely, *even today we don't know why the Pauli exclusion principle is true.*

Kitty said, "If I were living back in the 1920s, I would have announced the Kitty exclusion principle right after Click! Click! Bohr said that electrons could only go in orbits of n =1, 2, 3, etc. That would have stopped Sommerfeld, Goudsmit, and Uhlenbeck from inventing all those subshells and sub-subshells and those funny numbers ℓ, m_ℓ, and m_s."

Fred explained, "That's not the way things work. We can't *tell* nature what an atom is. We have to *ask* nature. We have to do the observations and experiments. Old Democritus didn't do any looking. He just declared that atoms had to be small, hard, indestructible, and that there were an infinite number of them.

"When we started looking, we found that atoms are not hard, they are not indestructible, and that there are only a finite number of them."

"I want the universe to be simple!" Kitty shouted. "It's too much—all those electron balls running around in fixed orbits and sub-orbits. It would take me a week to learn all that stuff."

---

* Bad example. Oops! Did I just write another sentence fragment? Social Security came into existence in the 1930s.

Fred asked a question that stopped Kitty: "What makes you think that human beings with their three-pound brains should be able to understand everything? There are mysteries in physics. There are mysteries in mathematics. There are mysteries in religion. Only lunatics and God say that they know all the secrets of the universe."

Kitty was silent.

"Let me continue," Fred said. "In two more years things get really spooky."

1926

You would think that after Pauli declared that those electron balls rolling around in those n, ℓ, m_ℓ, and m_s energy states was a complete and final description—you would think that that was enough. Physicists could close their laboratories and go on vacation.

The orbits were all determined: n, ℓ, m_ℓ, and m_s say it all.

Electron balls in known orbits—can you guess what is coming?

One year after Pauli seemed to finish the discussion of orbits and sub-orbits, Erwin Schrödinger worked on the "electron balls" idea. Three years earlier Louis de Broglie said that matter can be thought of as a wave. So Erwin wrote an equation of an electron as a wave. He called it Schrödinger's equation. His mother must have been very proud.

1927

Werner Heisenberg announced that we had lost the electron. He called it his **uncertainty principle**. He showed that we will never be able to pin down the exact position of an electron. We will never be able to determine the exact velocity of an electron.*

Heisenberg didn't just declare his uncertainty principle like Democritus might have. He demonstrated that we will never be able to locate an electron and say, "This is where it is and this is its velocity"—even with the best scientific equipment that can be bought on the Internet.

---

* To get technical for a moment, Heisenberg showed that the error in measuring the position multiplied by the error in measuring the mass times velocity will always be greater than or equal to $\hbar$.

(error in position)(error in mass times velocity) $\geq \hbar$.

$\hbar$ ("h-bar") is a tiny physics constant.

The electron has become fuzzy. It is a cloud, a fog, a haze. With Schrödinger's wave/matter equation, all that physicists can say is that there is a *probability* that an electron is at a particular point.

It is not a matter of physicists not knowing where that electron is. It is simply that they can *never* pin it down and say where it is or what its velocity is.

Worse yet, Heisenberg's uncertainty principle applies to any particle, not just electrons. At the atomic level, no human will ever be able to tell where things are, where they were in the past, or where they will be in the future. Everything is as messy as some teenagers' bedrooms.

Kitty stood up and said, "Thanks for the atom history, but we've got to get to work. The lunch crowd will be here soon."

Fred was disappointed. He had wanted to complete his story of the atom. In the time it took Kitty to finish off eight slices of bacon, two English muffins, and a quart of chocolate milkshake, Fred had not even gotten to the nucleus of an atom, other than saying that it was really small in comparison to the volume of the whole atom.

In fact, he hadn't even finished with the electron.

Schrödinger's equation was a mathematics equation. Yet it made all the same predictions as Bohr and Sommerfeld made. Schrödinger's equation generated n, ℓ, and m_ℓ. That was not bad for one equation.

1928

Two years later Paul Dirac created a new equation. It took into account Einstein's theories of relativity (1905 and 1915). This equation worked with many properties of matter: heat, crystals, electric circuits, and magnetism.

Dirac's equation also generated all four of the quantum numbers: n, ℓ, m_ℓ, and m_s. It predicted properties of electrons that were correct to within 0.001% accuracy. Paul's mother must have been really proud. We don't know what Erwin Schrödinger's mom felt about Paul Dirac's work. Maybe she had a bumper sticker that read: MY SON ERWIN HAD THE BEST ELECTRON EQUATION IN 1926–1927.

One strange thing about Dirac's 1928 equation was that it predicted something really weird. It said that brother electron had a sister who was just like him—except she had a positive charge instead of a negative one.

That must have made Dirac's equation look pretty stupid.

1932

Four years later, Carl David Anderson found the sister particle. He called it a positron. Paul Dirac's mother must have been very relieved.

That same year, James Chadwick found a cousin particle to the proton which is also in the nucleus of an atom. It was virtually identical to the positively charged proton, except that it had no charge. Since it was electrically neutral, it was called the neutron.

Four particles in atom: electrons, positrons, protons, and neutrons. All of these were found by the year 1932. Everything was now nice and neat. ☺ It might take Kitty a whole week to learn about these four particles if she were not busy running the Kitty Cafe. The happy thought is that by 1932 physicists had finally finished the story of the atom that began centuries ago with Democritus.

We won't spoil the story by mentioning the hundreds of particles that have been discovered since 1932.

The more we discover, the less we pretend to understand.

The Bridge
answers

from p. 38—*first try*

1. 23.4 pounds

$F = \mu N$ where $N = 78$ and $\mu = 0.3$

$F = (0.3)(78) = 23.4$ pounds. Joe would need to pull with a force of 23.4 pounds to keep the boat moving at a constant speed.

2. 56.4 lbs.

The boat weighs 78 lbs. Darlene weighs 110 lbs. Together they weigh 188 lbs. N = 188. $\mu = 0.3$

$F = \mu N = (0.3)(188) = 56.4$ lbs.

3. $\mu = 0.6$

$F = \mu N \qquad \frac{F}{N} = \mu \qquad \frac{46.8}{78} = 0.6$

$$\begin{array}{r} 0.6 \\ 78\overline{)46.8} \\ \underline{468} \end{array}$$

4. 1813 square centimeters

$A = lw = (37)(49) = 1813$

$$\begin{array}{r} 49 \\ \times\ 37 \\ \hline 343 \\ 147\ \\ \hline 1813 \end{array}$$

5. Yes. The number of fish Joe caught is a discrete variable. He could catch 4 or 5 fish, but he couldn't catch 4.0873 fish.

6. No. The weight of the first fish he caught is not a discrete variable. He could catch a 4-pound fish or a 5-pound fish or a fish that would weigh any amount between 4 and 5 pounds.

7. No. The clock will not tick slower. The period of a pendulum is independent of the weight of the pendulum.

8. $\frac{8}{9}$ pounds.

$F = \mu N \qquad N = 2\frac{2}{3} \qquad \mu = \frac{1}{3}$

$F = (2\frac{2}{3})(\frac{1}{3}) = \frac{8}{3} \times \frac{1}{3} = \frac{8}{9}$

> To convert $2\frac{2}{3}$ into an improper fraction multiply 3 times 2 and add it to the 2 in the numerator.

9. Yes. Numerals are the written objects.

10. $\mu = \frac{8}{21}$

$8 = 21\mu$

Divide both sides by 21 $\qquad \frac{8}{21} = \frac{21\mu}{21}$

Simplify $\qquad \frac{8}{21} = \mu$

The Bridge
answers

from p. 39—*second try*

1. 24 lbs.

$F = \mu N \qquad \mu = 0.4 \qquad N = 60$ lbs.

$F = (0.4)(60) = 24$

2. 23.2 lbs.

The cake originally weighed 60 lbs. When the 2-pound doll fell off it weighed 58 lbs.

$F = \mu N = (0.4)(58) = 23.2$ lbs.

3. $\mu = 0.2$ or $\mu = \frac{1}{5}$

$F = \mu N \qquad \frac{F}{N} = \mu \qquad \mu = \frac{F}{N} = \frac{34}{170} = \frac{1}{5}$ or 0.2.

4. 93.5 square inches or $93\frac{1}{2}$ square inches

$A = lw = 8\frac{1}{2} \times 11 = \frac{17}{2} \times \frac{11}{1} = \frac{187}{2} = 93\frac{1}{2}$

5. No. Her dress could cost \$300.05 or \$300.06 but it couldn't cost any amount between 300.05 and 300.06.

6. No. She could have 2 bridal attendants or 3 of them, but she couldn't have 2.7 of them.

7. $6\frac{1}{4}$ feet per second or 6.25 feet per second

$d = rt \qquad \frac{d}{t} = r \qquad r = \frac{d}{t} = \frac{50}{8} = 6\frac{1}{4}$ or 6.25

8. $\mu = \frac{9}{14}$

$14\mu = 9$

Divide both sides by 14 $\qquad \frac{14\mu}{14} = \frac{9}{14}$

Simplify $\qquad \mu = \frac{9}{14}$

9. $26\frac{2}{3}$ hours or 26 hours, 40 minutes

$d = rt \qquad \frac{d}{r} = t \qquad t = \frac{d}{r} = \frac{1600}{60} = 26\frac{2}{3}$

Two-thirds of an hour $= \frac{2}{3} \times 60$ minutes $= 40$ minutes

10. 0.5 or $\frac{1}{2}$

$2.5 = 5x$

Divide both sides by 5 $\qquad \frac{2.5}{5} = \frac{5x}{5}$

Simplify $\qquad 0.5 = x$

The Bridge
answers

from p. 40—*third try*

1. $\mu = \frac{1}{3}$

$$F = \mu N \qquad \frac{F}{N} = \mu \qquad \mu = \frac{1/3}{1} = \frac{1}{3}$$

2. 36 inches

$$d = rt = (3)(12) = 36$$

3. 1.14 ounces

$$F = \mu N = (0.19)(6) = 1.14$$

$$\begin{array}{r} 0.19 \\ \times \quad 6 \\ \hline 114 \end{array} \quad 1.14$$

4. $27\frac{27}{32}$ square inches

$$A = lw = 4\frac{1}{8} \times 6\frac{3}{4} = \frac{33}{8} \times \frac{27}{4} = \frac{891}{32} = 27\frac{27}{32}$$

$$\begin{array}{r} 27 \text{ R } 27 \\ 32\overline{)891} \\ \underline{64} \\ 251 \\ \underline{224} \\ 27 \end{array}$$

5. Yes. He could own 27 of her books or 28 of them, but he couldn't own $27\frac{3}{8}$ of them.

6. No. He could press with 6 ounces of pressure or 7 ounces or any number between 6 and 7.

7. $\mu = \frac{100}{77}$ or $1\frac{23}{77}$

$$77\mu = 100$$

Divide both sides by 77 $\quad \frac{77\mu}{77} = \frac{100}{77}$

Simplify $\quad \mu = \frac{100}{77}$

8. 77 is a larger number than 8.

9. $x = \frac{8}{43}$

$$8 = 43x$$

Divide both sides by 43 $\quad \frac{8}{43} = \frac{43x}{43}$

Simplify $\quad \frac{8}{43} = x$

10. The 5-foot blue thread has a longer period. The weight does not affect the period. Longer pendulums have longer periods.

The Bridge
answers

from p. 41—*fourth try*

1. $\mu = \frac{1}{3}$

$F = \mu N \qquad \frac{F}{N} = \mu \qquad \mu = \frac{50}{150} = \frac{1}{3}$

2. $44\frac{2}{3}$ pounds

The bag now weighs 134 pounds (150 – 16). $\mu = \frac{1}{3}$

$F = \mu N = \frac{1}{3} \times 134 = \frac{1}{3} \times \frac{134}{1} = \frac{134}{3} = 44\frac{2}{3}$

3. No. You can have 3 bowling balls or 4 bowling balls, but you can't have 3.897 bowling balls.

4. Yes. Jack could weigh 152 or 152.14 pounds.

5. 2.5 miles per hour or $2\frac{1}{2}$ miles per hour

$d = rt \qquad \frac{d}{t} = r \qquad r = \frac{d}{t} = \frac{15}{6} = 2\frac{3}{6} = 2\frac{1}{2}$

6. 19,500 square yards

$A = lw = 150 \times 130 = 19{,}500$

7. 7 seconds

The period of a pendulum is independent of the weight. Only the length matters. Since the length did not change, the period did not change.

8. x = 20

$4 = 0.2x$

Divide both sides by 0.2 $\qquad \frac{4}{0.2} = \frac{0.2x}{0.2}$

Simplify $\qquad 20 = x$

$0.2\overline{)\,4.}$

becomes $2.\overline{)\,40.}$

9. Your answer may be different than mine.

6 and 8	6 is the larger numeral and the smaller number.
3 and 9	3 is the larger numeral and the smaller number.
5 and 6	5 is the larger numeral and the smaller number.

10. $x = \frac{18}{25}$

$4\frac{1}{2} = (6\frac{1}{4})x$

Divide both sides by $6\frac{1}{4}$ $\qquad \frac{4\frac{1}{2}}{6\frac{1}{4}} = x$

$4\frac{1}{2} \div 6\frac{1}{4} = \frac{9}{2} \div \frac{25}{4} = \frac{9}{2} \times \frac{4}{25} = \frac{36}{50} = \frac{18}{25}$

The Bridge
answers

from p. 42—*fifth try*

1. $\mu = \frac{1}{4}$ or 0.25

$F = \mu N \qquad \frac{F}{N} = \mu \qquad \mu = \frac{9}{36} = \frac{1}{4}$

2. $12\frac{1}{2}$ pounds or 12.5 pounds

Adding 14 pounds of extra cheese made the pizza weigh 50 pounds. $\mu = \frac{1}{4} \qquad F = \mu N = (\frac{1}{4})(50) = \frac{50}{4} = 12\frac{2}{4} = 12\frac{1}{2}$

3. 1,488 square inches

$A = \ell w = 48 \times 31 = 1{,}488$

4. Neither. The period would remain the same. The period of a pendulum does not depend on the weight.

5. $\mu = \frac{1}{2}$ or 0.5

$30\mu = 15$

Divide both sides by 30 $\qquad \frac{30\mu}{30} = \frac{15}{30}$

Simplify $\qquad \mu = \frac{1}{2}$

6. Yes. If a pizza can weigh 17.3 pounds or 17.4 pounds, it can weigh 17.382 pounds.

7. No. He could make 17 pizzas or 18 pizzas, but he couldn't make 17.382 pizzas.

8. **5** is a larger numeral than 9.

9. $\mu = \frac{1}{3}$

$F = \mu N \qquad \frac{F}{N} = \mu \qquad \mu = \frac{5\frac{1}{3}}{16} = \frac{1}{3}$

The details: $5\frac{1}{3} \div 16$

$= \frac{16}{3} \div \frac{16}{1}$

$= \frac{16}{3} \times \frac{1}{16}$

$= \frac{\cancel{16}}{3} \times \frac{1}{\cancel{16}}$

$= \frac{1}{3}$

10. $y = \frac{5}{7}$

$7y = 5 \quad \Rightarrow \quad \frac{7y}{7} = \frac{5}{7} \quad \Rightarrow \quad y = \frac{5}{7}$

The Bridge

answers

from p. 67—*first try*

1. Inductive reasoning. By catching a lot of fish, Joe had done many experiments. Each experiment told him that the particular fish he had caught was gray. He generalized from those observations that all the fish in the Great Lake are gray. His conclusion, like all scientific conclusions, was only probably true. The fish that he catches tomorrow may be a goldfish.

2. Hooke's law is F = kx—the force applied to a spring is proportional to the amount of stretch of the spring. (This is only true within the proportional limits of the spring.)

3. Start with $77 = 13\mu$

Divide both sides by 13 $\frac{77}{13} = \frac{13\mu}{13}$

Simplify both sides $5\frac{12}{13} = \mu$

4. 23 minutes. We are given 345 meters and 15 meters/minute. If you don't know whether to add, subtract, multiply, or divide, restate the problem using simpler numbers. Suppose we had 6 meters and were winding at the rate of 2 meters/minute. It would take 3 minutes. We divided. In the original problem: $15\overline{)\ 345}$

5. Within either the proportional or the elastic limits, the fishing line would have returned to its normal shape after he caught the fish. If the fish pulled hard enough so that the line entered the plastic region, the fish would still be caught, although the line would be permanently stretched. What Joe should be worried about is D) the breaking point. If the fish broke the line, it would escape.

6. Hooke's law is F = kx. A force of 5 ounces will stretch the spring $3\frac{1}{3}$ inches. $5 = k(3\frac{1}{3})$.

Putting the number in front of the letters $5 = (3\frac{1}{3})k$

Dividing both sides by $3\frac{1}{3}$ $\frac{5}{(3\ 1/3)} = k$

Doing the arithmetic $5 \div 3\frac{1}{3} = 5 \div \frac{10}{3} = \frac{5}{1} \times \frac{3}{10} = \frac{5}{1} \times \frac{3}{10} = \frac{3}{2}$

$k = \frac{3}{2}$ or $1\frac{1}{2}$ or 1.5.

The Bridge
answers

7. $F = kx$ where $x = 2"$ and, from the previous problem, $k = \frac{3}{2}$

$F = \frac{3}{2} \times 2 = 3$ ounces.

8. No. The number of birds is a discrete variable. You can have six birds or seven birds, but you can't have $6\frac{1}{5}$ birds.

9. Area of a rectangle is equal to length times width. $A = \ell w$.
$A = 8 \times 10.5 = 84$ square inches.

10. The motion of the line was being transformed into heat. (The energy of motion is sometimes called kinetic energy.)

From p. 69—*second try*

1. $A = \ell w = (14)(23) = 322$ square inches.

$$\begin{array}{r} 23 \\ \times 14 \\ \hline 92 \\ 23 \\ \hline 322 \end{array}$$

2. $\mu = \frac{2}{3}$

$$F = \mu N$$
$$\frac{F}{N} = \mu$$
$$\frac{28}{42} = \mu \qquad \text{and } \frac{28}{42} \text{ reduces to } \frac{2}{3}$$

3. $\frac{4}{5}$

$$24 = 30x$$

Divide both sides by 30 $\quad \frac{24}{30} = \frac{30x}{30}$

Simplify both sides $\quad \frac{4}{5} = x \quad$ (or 0.8)

4. $\frac{1}{3}$

$$F = \mu_s N$$
$$4 = \mu_s 12$$

Putting the number in front of the letter $\quad 4 = 12\mu_s$

Divide both sides by 12 $\quad \frac{1}{3} = \mu_s$

5. The energy associated with height was converted into motion which was then converted into sound.

6. Speed is a continuous variable. At one moment the magazine might be falling at 2 m/sec (meters per second) and at another moment it might be falling at 2.989262005 m/sec.

7. 0.5 inches.

Hooke's law $\quad F = kx$

Given $F = 4.8$ and $x = 3$ $\quad 4.8 = 3k$

Divide both sides by 3 $\quad 1.6 = k$

The Bridge
answers

So in this case Hooke's law is	$F = 1.6x$
For F = 0.8	$0.8 = 1.6x$
Dividing both sides by 1.6	$\frac{0.8}{1.6} = \frac{1.6x}{1.6}$
Simplifying both sides	$0.5 = x$

8.

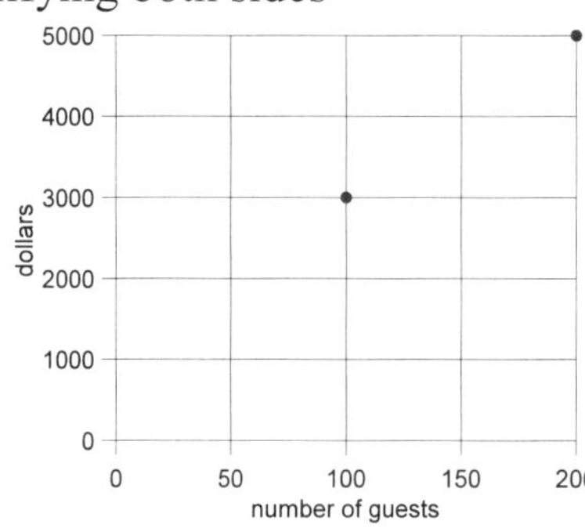

9. No. Without actually doing an experiment, you cannot find the value of μ_k. All we know is that μ_k is probably less than μ_s.

10. $\frac{A}{w}$ $\qquad A = \ell w$

Divide both sides by w $\qquad \frac{A}{w} = \frac{\ell w}{w}$

Simplifying $\qquad \frac{A}{w} = \ell$

from p. 70—*third try*

1. $\frac{1}{3}$ $\qquad F = \mu_k N$

$5 = \mu_k(15)$

Divide both sides by 15 $\qquad \frac{5}{15} = \mu_k$

Simplify $\qquad \frac{1}{3} = \mu_k$

2. 5⅔ pounds $\qquad F = \mu_k N$

$F = \frac{1}{3}(17)$

Simplifying $\qquad F = 5\frac{2}{3}$

3. 36.4 sq cm $\qquad A = \ell w = (5.2)(7) = 36.4$ sq cm

4. No. You cannot have, for example, 3.08987 alligators.

The Bridge
answers

5. 38 feet/minute

$$d = rt$$
$$76 = r(2)$$
Divide both sides by 2 $\quad 38 = r$

6. *1* is a larger numeral than *8*.

7. $6\frac{1}{4}$ inches

This is a two-step problem. We first find k.

Hooke's law $\quad F = kx$

Given F = 16 and x = 10 $\quad 16 = k(10)$

Divide both sides by 10 $\quad 1.6 = k$

Second step.

Given F = 10 and from the first step, k = 1.6

F = kx becomes $\quad 10 = 1.6x$

Divide both sides by 1.6 $\quad \frac{10}{1.6} = x$

Simplify $\quad 6.25 = x \quad (\text{or } 6\frac{1}{4})$

8. Inductive reasoning.

Fred went from experiments/trials/observations to a general conclusion. This is the way that scientists often operate. The conclusions of inductive reasoning are always probable and not absolutely certain.

9. $\frac{3}{80}$ $\quad 80y = 3$

Divide both sides by 80 $\quad y = \frac{3}{80}$

10. c is exactly 299,792,458 m/sec, not because we have measured the speed of light, but because we have defined a meter to be exactly the distance that light can travel in $\frac{1}{299{,}792{,}458}$ of one second.

from p. 72—*fourth try*

1. 1.25 square miles. The area of a rectangle is $A = \ell w$. (0.5)(2.5) = 1.25
2. Jack's period was also equal to three seconds.

The period of a pendulum is independent of the weight. Only the length matters. Since the length did not change, the period did not change.

3. μ_k This is the coefficient for kinetic (motion) friction. As long as he was sliding, μ_s did not apply.
4. x is equal to 4. $\quad 52 = 13x$

Divide both sides by 13 $\quad \frac{52}{13} = x$

$$13\overline{)52}$$ quotient 4, subtract 52

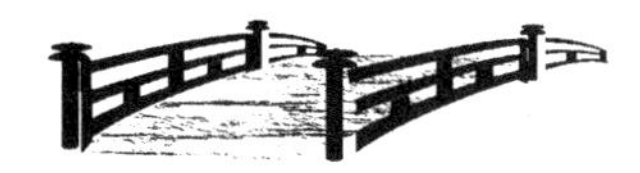

The Bridge
answers

5. 36 lbs.

This is a two-step problem. First we find μ_s.

Given $N_{Jake} = 155$ and $F = 31$, $F = \mu_s N$ becomes $31 = \mu_s(155)$

Putting the number in front of the letter $31 = 155\mu_s$

Dividing both sides by 155 $1/5 = \mu_s$

Secondly, we find F_{Jack}.

Given $N_{Jack} = 180$ and $\mu_s = 5$, $F = \mu_s N$ becomes $F = (1/5)180$

$F = 36$

6. Yes. The number of films is a discrete variable. If the number of films were a continuous variable, then it would make sense to talk about 305.9392 films shot in Central Park.

7. Jane's pay is a discrete variable. She might be paid \$96.55 or \$96.56, but she couldn't be paid \$96.5503.

8. When Jake rubbed the green stains with his handkerchief, he was converting kinetic energy into heat. That is what friction does.

9. Her reasoning was inductive. She was moving from observations/trials/experiments to a general conclusion (that rubbing grass stains with a handkerchief will not work). Her conclusion could only be probably true, not absolutely true. It might turn out that with a certain type of grass stain, a particular kind of handkerchief, and more forceful rubbing, the stains might easily disappear. Every conclusion of science is tentative.

10. $x = 9$

$(\frac{2}{3})x = 6$

Divide both sides by $\frac{2}{3}$ $x = 9$

The details from arithmetic: $6 \div \frac{2}{3} = \frac{6}{1} \times \frac{3}{2} = \frac{18}{2} = 9$

from p. 74—*fifth try*

1. Inductive reasoning. Stanthony went from trials/observations/ experiments to his general conclusion that he should never use rice as a pizza topping.

2. The pizza flipper had entered the plastic region. If it had been within the proportional or elastic limits, it would have sprung back to its original shape. If it had passed the breaking point, it would have broken into two pieces.

The Bridge
answers

3. $y = \frac{4}{3}$ (or $1\frac{1}{3}$)

$$18y = 24$$

Divide both sides by 18 $\qquad \frac{18y}{18} = \frac{24}{18}$

Simplify both sides $\qquad y = \frac{4}{3}$

Arithmetic details: $\frac{24}{18}$ reduces to $\frac{4}{3}$ when you divide top and bottom by 6.

4. 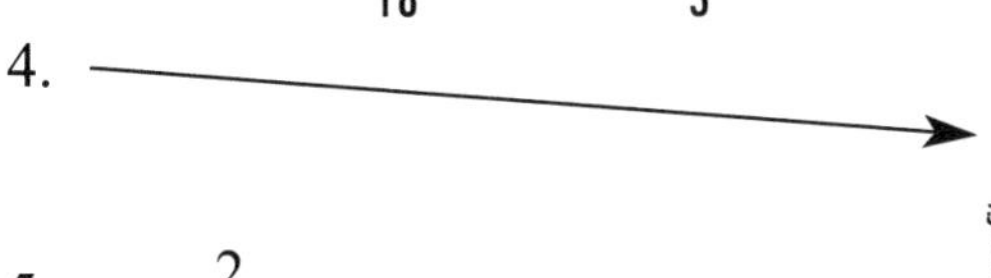

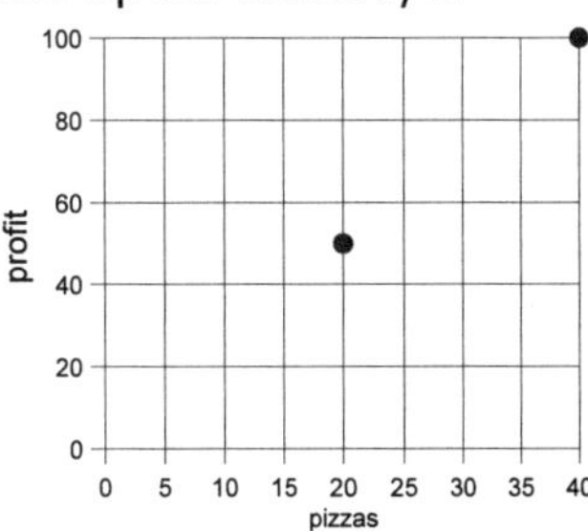

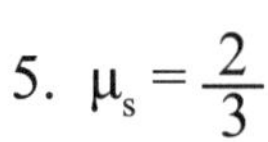

5. $\mu_s = \frac{2}{3}$

Given F = 40 and N = 60.

$F = \mu_s N$ becomes $\qquad 40 = \mu_s(60)$

Divide both sides by 60 $\qquad \frac{2}{3} = \mu_s$

Arithmetic details: $\frac{40}{60}$ reduces to $\frac{2}{3}$ when you divide top and bottom by 20.

6. No. The coefficient of sliding friction (μ_k) cannot be determined from μ_s. It has to be determined by experiment.

7. 3,500 square inches.

$$A = lw = (50)(70) = 3{,}500$$

8. 5.6 pounds

$$F = \mu_k N = (0.7)(8) = 5.6$$

9. No. The number of customers is a discrete variable. It can be 56 or 57, but it cannot be 56.00038.

10. Eight is a larger number than three.

from p. 101—*first try*

1. $F = \mu_k N$ where $\mu_k = 0.25$ and N = 210 lbs.
   F = (0.25)(210) = 52.5 lbs.
   Since Joe is pulling with a force of 60 lbs. the boat will accelerate.

2. To pull at a constant speed, he would use a force of $F = \mu_k N =$ (0.25)(210) = 52.5 lbs. W = Fd = (52.5)(100) = 5,250 ft-lb

3. $F = \mu_s N$ $\qquad$ Divide both sides of the equation by N to get $\frac{F}{N} = \mu_s$

$$\mu_s = \frac{F}{N} = \frac{63}{210} = 0.3$$

$$210\overline{)63.0} = 0.3$$
$$\underline{630}$$

The Bridge
answers

4.

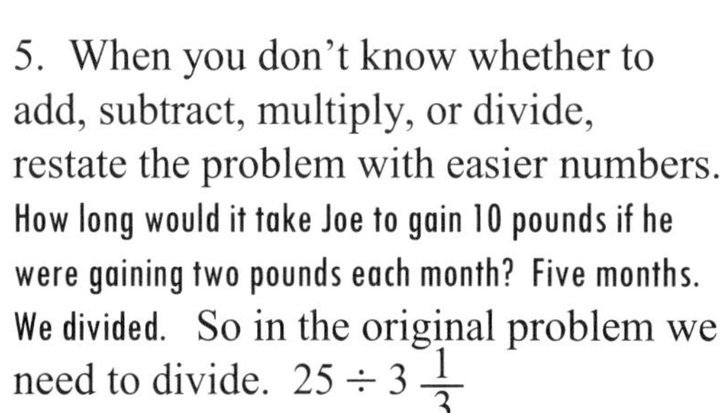

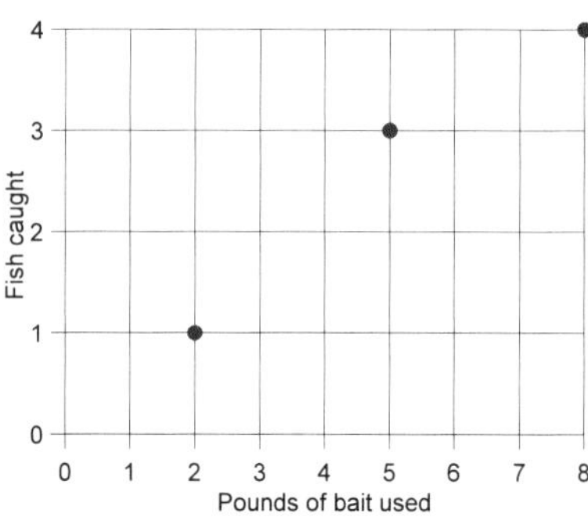

5. When you don't know whether to add, subtract, multiply, or divide, restate the problem with easier numbers. How long would it take Joe to gain 10 pounds if he were gaining two pounds each month? Five months. We divided. So in the original problem we need to divide. $25 \div 3\frac{1}{3}$

$\frac{25}{1} \div \frac{10}{3}$ To convert $3\frac{1}{3}$ into an improper fraction you say "3 times 3 plus 1."

$\frac{25}{1} \times \frac{3}{10} = \frac{75}{10} = 7.5$ months (or $7\frac{1}{2}$ months).

6. $\frac{6 \text{ tablespoons}}{1} \times \frac{100 \text{ Calories}}{1 \text{ tablespoon}} = 600$ Calories

$\frac{600 \text{ Calories}}{1} \times \frac{3088 \text{ ft-lb}}{1 \text{ Calorie}} = 1{,}852{,}800$ ft-lb

7. $36y = 75$

$\frac{36y}{36} = \frac{75}{36}$ Divide both sides by 36

$y = \frac{75}{36} = 2\frac{3}{36} = 2\frac{1}{12}$ Simplify

8. 53°. The first two angles, 49° and 78°, add to 127°. Since all three angles of any triangle add to 180°, that leaves 53° (180° – 127°) for the third angle.

9. This is an example of deductive reasoning. Joe was moving from given information (that one fishhook costs 30¢) to what must be true because of that given information.

10. In 1983, the Seventeenth General Conference on Weights and Measures used the speed of light to declare that a meter was the length that light travels in $\frac{1}{299{,}792{,}458}$ of a second. The speed of light was used to establish the length of a meter.

The Bridge
answers

from p. 102—*second try*

1. The number of bridesmaids is a discrete variable. You can have two or three bridesmaids, but you can't have 2.9863 bridesmaids.

2. One table = $5,000, so the conversion factor is either $\frac{1\text{ table}}{\$5000}$ or it is $\frac{\$5000}{1\text{ table}}$ We choose the one that will make the units cancel.

$$\frac{20\text{ tables}}{1} \times \frac{\$5000}{1\text{ table}} = \$100{,}000$$

3. 0.6

$$F = \mu_s N \qquad \frac{F}{N} = \mu_s$$

$$\mu_s = \frac{F}{N} = \frac{180}{300} = 0.6$$

$$\begin{array}{r} 0.6 \\ 300\overline{)\,180.0} \\ \underline{1800} \end{array}$$

4. 150 lbs.

$$F = \mu_s N \qquad \frac{F}{\mu_s} = N$$

$$N = \frac{F}{\mu_s} = \frac{120}{0.8} = 150$$

$0.8\overline{)\;120.}$ becomes

$$\begin{array}{r} 150 \\ 8\overline{)\,1200} \\ \underline{8} \\ 40 \\ \underline{40} \end{array}$$

5. Inductive reasoning. Queen Boodle did observations/experiments/trials by looking at the first 80 names on the list. From there she generalized that all 432 names were former boyfriends.

6. 5 minutes. Given 200 gallons and 40 gallons per minute, the question is whether to add, subtract, multiply, or divide. When you are not sure, restate the problem with easier numbers. How long would it take to fill a 6-gallon tub at 2 gallons per minute? Three minutes. We divided.

In the original problem we divide: $40\overline{)\,200}$ = 5

7. $\frac{1}{4}$ or 0.25 μ_s is defined as $\frac{F}{N}$ and we have shown this is equal to $\frac{\text{Rise}}{\text{Run}}$ of the surface. $\frac{\text{Rise}}{\text{Run}} = \frac{3}{12} = \frac{1}{4}$ or 0.25

8. It is not possible. The only thing we know is that $\mu_k \le \mu_s$.

9.

3 lbs.

3"

W

12"

It does not matter how long you draw this vector arrow. Once you have drawn that vector, then resolving that vector into N and F will give you the relative lengths of N and F with respect to the weight vector W. It is the ratios between W, N, and F that matter.

10. Continuing the drawing made in the previous problem.

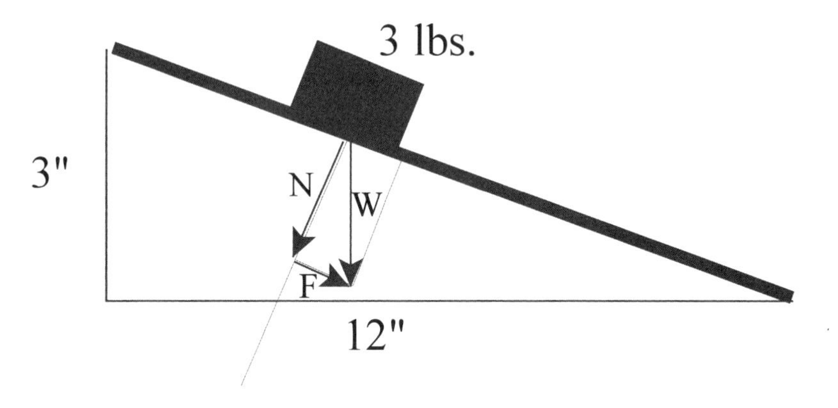

Just for fun, I measured the lengths of N and W in my drawing and found that N was 97% as long as W. So the normal force N is 97% of the 3-pound weight. N is approximately 2.9 pounds.

from p. 104—*third try*

1. Since 3 tears = 8 lines, the conversion factor will be either $\frac{3 \text{ tears}}{8 \text{ lines}}$ or it will be $\frac{8 \text{ lines}}{3 \text{ tears}}$

We are starting with 112 lines, so we pick the conversion factor so that the "lines" will cancel.

$$\frac{112 \text{ lines}}{1} \times \frac{3 \text{ tears}}{8 \text{ lines}} = \frac{336}{8} \text{ tears} = 42 \text{ tears.}$$

The Bridge
answers

2. 320 square inches. The area of a rectangle ($A = \ell w$) is equal the length times the width. $A = (20)(16) = 320$

3. 0.5 pounds. $F = \mu_k N$ The force is 0.3 pounds and μ_k is 0.6

$$0.3 = 0.6N$$

Divide both sides by 0.6 $\quad \frac{0.3}{0.6} = \frac{0.6N}{0.6}$

$0.6\overline{)0.3}$ becomes $6.\overline{)3.0}$ = 0.5 (3.0 − 30)

Simplify both sides $\quad 0.5 = N$

4. 5 hours. He reads 60 pages at the rate of 12 pages per hour. If you don't know whether to add, subtract, multiply, or divide, restate the problem with easier numbers and notice which operation you use. If he read 6 pages at the rate of 3 pages per hour, it would take him 2 hours. Division was used. 6 pages divided by 3 pages per hour.

So 60 pages divided by 12 pages per hour = 5 hours.

5. Yes. This is what Kingie taught Fred in Chapter 13. He told Fred to just measure the Rise and the Run in an experiment.

6. Given $\quad d = rt$

Divide both sides by r $\quad \frac{d}{r} = \frac{rt}{r}$

Simplify $\quad \frac{d}{r} = t$

7.

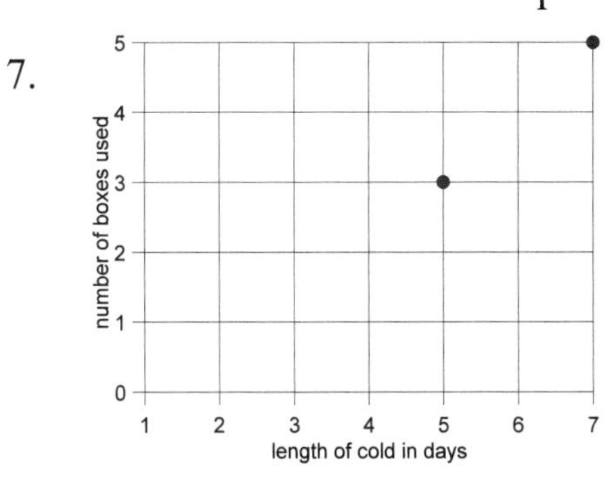

8. The largest numeral is 3.

9. The largest number is 800.

10. The largest digit is 8. (The other digits are 0 and 3.)

The Bridge
answers

from p. 106—*fourth try*

1. $4\frac{1}{2}$ hours.

How long will take to gain $11\frac{1}{4}$ pounds if you are gaining $2\frac{1}{2}$ pounds per hour? When you don't know whether to add, subtract, multiply, or divide, restate the problem using simpler numbers. How long will take to gain 6 pounds if you are gaining 2 pounds per hour? It will take 3 hours. We divided 6 by 2.

In the original problem we divide $11\frac{1}{4}$ by $2\frac{1}{2}$

$$11\frac{1}{4} \div 2\frac{1}{2} = \frac{45}{4} \div \frac{5}{2} = \frac{45}{4} \times \frac{2}{5} = \frac{\overset{9}{\cancel{45}}}{\underset{2}{\cancel{4}}} \times \frac{\overset{1}{\cancel{2}}}{\underset{1}{\cancel{5}}} = \frac{9}{2} = 4\frac{1}{2}$$

2. 0.4

$F = \mu_k N$	$64 = \mu_k(160)$
Divide both sides by 160	$\frac{64}{160} = \frac{\cancel{\mu_k}160}{160}$
Simplify	$0.4 = \mu_k$

$$160\overline{)64.0}\quad 0.4,\quad \underline{640}$$

3. 30º

There are 180º in a triangle. 105º + 45º = 150º. So the third angle must be 30º.

4. Inductive reasoning. Jack started with observations/trials/experiments and arrived at the conclusion that there would be a cell phone ad on every page of the magazine. Inductive reasoning leads to conclusions that are only probably true, rather than absolutely true.

5. $\mu_s = 0.6$

Kingie has shown that μ_s could be computed by $\frac{\text{Rise}}{\text{Run}}$ (which is the slope of the line). In this case, $\frac{12}{20} = \frac{6}{10} = 0.6$ (or $\frac{3}{5}$)

The Bridge
answers

6. Since 24 minutes equals \$16, the conversion factor will either be $\frac{24 \text{ minutes}}{\$16}$ or it will be $\frac{\$16}{24 \text{ minutes}}$

We want to convert \$56 into minutes so we select the first conversion factor so that the units will cancel.

$$\frac{\cancel{\$}56}{1} \times \frac{24 \text{ minutes}}{\cancel{\$}16} = \frac{56}{1} \times \frac{\overset{3}{\cancel{24}} \text{ minutes}}{\underset{2}{\cancel{16}}} = 84 \text{ minutes}$$

7. Use the other conversion factor so that the minutes will cancel.

$$\frac{90 \cancel{\text{minutes}}}{1} \times \frac{\$16}{24 \cancel{\text{minutes}}} = \frac{90}{1} \times \frac{\$\overset{2}{\cancel{16}}}{\underset{3}{\cancel{24}}} = \$60$$

8. $8x = 3$

Divide both sides by 8 $x = \frac{3}{8}$

9. The number of pigeons is a discrete variable. You cannot have 8.39 pigeons.

10. A motor changes electrical energy into kinetic (or motion) energy.

from p. 108—*fifth try*

1. 81 ft-lb

$W = Fd = (27)(3) = 81$

2. 0 ft-lb

In physics you have to move an object in order to perform work. If d = 0, then Fd is equal to zero.

3. 10.8 lbs.

$F = \mu_k N = (0.4)(27) = 10.8$

4. 36 minutes

Since 18 pizzas equals 27 minutes, the conversion factor will either be $\frac{18 \text{ pizzas}}{27 \text{ minutes}}$ or $\frac{27 \text{ minutes}}{18 \text{ pizzas}}$

We choose the one that makes the units cancel.

$$\frac{24 \cancel{\text{pizzas}}}{1} \times \frac{27 \text{ minutes}}{18 \cancel{\text{pizzas}}} = \frac{24}{1} \times \frac{27 \text{ minutes}}{18} = 36 \text{ minutes}$$

The Bridge
answers

5. $\mu_s = \frac{1}{7}$

$F = \mu_s N$

$$11 = \mu_s(77)$$

Divide both sides by 77 $\qquad \frac{11}{77} = \mu_s$

$\frac{11}{77}$ reduces to $\frac{1}{7}$ when you divide top and bottom by 11.

6. The volume of starfish oil is a continuous variable. You can have 3 ounces of oil or 4 ounces or 3.5 ounces or 3.008236 ounces.

7. $\qquad 6x = 12$

Divide both sides by 6 $\qquad x = 2$

8. The Law of Conservation of Energy states that in any closed system, the amount of energy cannot change.

9. Friction changes the energy of motion (also known as kinetic energy) into heat.

10.

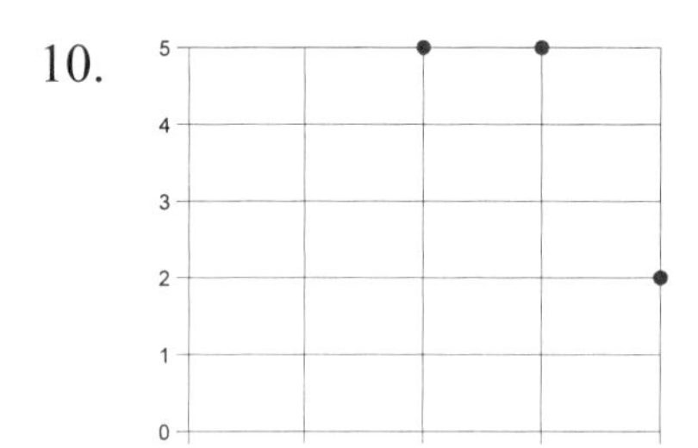

from p. 133—*first try*

1. Pressure at any point in a fluid is equal to the density of the fluid times the height of the surface above that point. $p = dh$

The end of the hose is 50 feet below the surface of the water. The pressure at that point is equal to (density)(height), which is (62 lbs./ft^3)(50 ft.) = 3100 lbs./ft^2

2. Draw a picture. Here is one square yard divided into square feet. One square yard = 9 square feet.

1	2	3
4	5	6
7	8	9

1 yard

1 yard

3. $4^3 = 4 \times 4 \times 4 = 16 \times 4 = 64$ Exponents were taught at the end of Chapter 23 after the *Your Turn to Play*.

The Bridge
answers

4. μ_s = Rise/Run = $\frac{1}{3}$

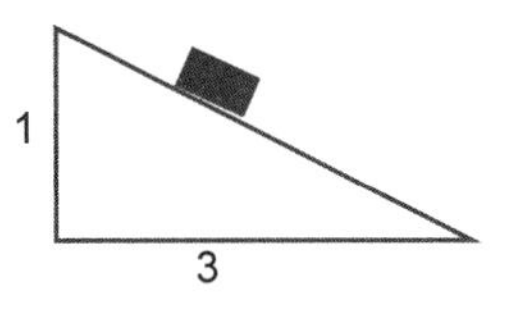

5. All we know is that μ_k is less than μ_s. We can't compute the coefficient of sliding friction from the given information.

6. $\sqrt{64} = 8$. (See the end of Chapter 3, after the *Your Turn to Play*.)

7. Since 24 square inches is equivalent to 9 minutes, the conversion factor will either be $\frac{24 \text{ in}^2}{9 \text{ min}}$ or it will be $\frac{9 \text{ min}}{24 \text{ in}^2}$

We start with 16 square inches and choose the conversion factor that will cancel the square inches.

$$\frac{16 \text{ in}^2}{1} \times \frac{9 \text{ min}}{24 \text{ in}^2} = \frac{16 \cancel{\text{in}^2}}{1} \times \frac{9 \text{ min}}{24 \cancel{\text{in}^2}} = \frac{\cancel{16}^{\,2}}{1} \times \frac{9 \text{ min}}{\cancel{24}_{\,3}}$$

= 6 minutes

8. Friction changes kinetic energy into heat.

9. 6 kilograms. Kilograms is a measure of mass. It measures how much "stuff" is in the can. That will not change. (The weight changes, but not the mass.)

10. Four fish caught corresponds to two hours.

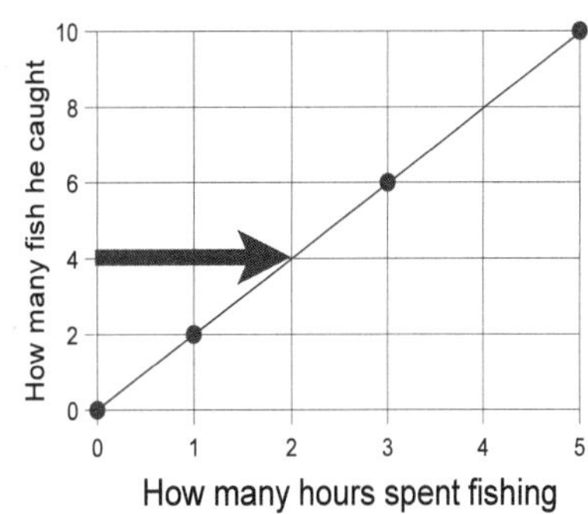

from p. 135—*second try*

1. Since 1 $\text{ft}^3 \approx 7.5$ gallons, the conversion factor will either be

$\frac{1 \text{ ft}^3}{7.5 \text{ gal.}}$ or it will be $\frac{7.5 \text{ gal.}}{1 \text{ ft}^3}$

The second conversion factor will either be $\frac{8 \text{ pints}}{1 \text{ gal.}}$ or it will be $\frac{1 \text{ gal.}}{8 \text{ pints}}$

We start with 1 ft^3 and choose the conversion factors that will make the units cancel.

$$\frac{1 \cancel{\text{ft}^3}}{1} \times \frac{7.5 \cancel{\text{gal.}}}{1 \cancel{\text{ft}^3}} \times \frac{8 \text{ pints}}{1 \cancel{\text{gal.}}} = 60 \text{ pints}$$

$$\begin{array}{r} 7.5 \\ \times \ \ 8 \\ \hline 600 \end{array} \rightarrow 60.0$$

(A note: Since a cubic foot of water weighs about 62 pounds, a cubic foot of bubbly water might weigh about 60 pounds. Then a pint would weigh a pound. Is that a coincidence?)

The Bridge
answers

2. 124 pounds/square foot. The pressure under a fluid is equal to the density times the depth = (62 lbs./ft^3)(2 ft) = 124 lbs./ft^2

3. 900 minutes. You are given 9,000 gallons and 10 gallons/minute. If you are not sure whether to add, subtract, multiply, or divide, restate the problem using simpler numbers: Suppose the pool was 6 gallons and you could fill it at the rate of 2 gallons per minute. That would take 3 minutes. You divided the volume of the pool by the rate of filling.

9,000 gallons divided by 10 gallons/minute = 900 minutes.
(That is 15 hours: $\frac{900 \text{ minutes}}{1} \times \frac{1 \text{ hour}}{60 \text{ minutes}} = 15$ hours.)

4. Inductive reasoning. Darlene had gone from observations/trials/experiments to a conclusion. Her conclusion is only probably true.

(On the other hand, if she was given the fact that all goldfish weigh less than 60 pounds and the fact that Roger is a goldfish, then she could conclude by deductive reasoning that Roger weighs less than 60 pounds. That conclusion would be certain—not just probably true.)

5. Area of a rectangle = (length)(width) = (20)(30) = 600 m^2.

6. 119 pounds $\quad F = \mu_k N = (0.7)(170) = 119$

7. 480 psi $\quad$ Pressure is defined as $\frac{\text{weight}}{\text{area}}$

The area of a square that is one-half inch on each side is $\frac{1}{4}$ square inches. Pressure = $120 \div \frac{1}{4} = 120 \times 4 = 480$

8. $\quad 12y = 7$

Divide both sides by 12 $\quad \frac{12y}{12} = \frac{7}{12}$

Simplify $\quad y = \frac{7}{12}$

9. The amount of water is a continuous variable. It could be 9,000 gallons or 9,000.0383 gallons.

10. There are two ways to determine the mass of an object. When there is no gravity, you cannot weigh the object. You have to measure its inertia. (If you mentioned the word *inertia*, you get full credit for this problem.) To measure mass, measure inertia. To measure inertia measure the force needed to change an object's speed or direction. Playing catch with the brick would tell you how much mass it had. Or just shaking the brick in your hand would give you an indication of how much mass it had.

The Bridge
answers

from p. 137—*third try*

1. Yes. If something on earth had a density of, say, 12 lbs./ft^3, it would have a density of 2 lbs./ft^3 on the moon. If you were swimming 10 feet under the surface of the water, the pressure you would feel on your ears would be one-sixth as much on the moon as on the earth.

What doesn't change is the mass. The volume of a liquid doesn't change. The length of a meter doesn't change. The length of a second doesn't change. In contrast, the weight of an object is directly related to gravity.

2. 4.8 cubic inches.

Cavalieri's Principle: volume = area of one of the poker chips times the height. V = (area of the base)(height). V = (0.8)(6) = 4.8 in^3.

3. First you draw the normal line, which is the line perpendicular to the surface.

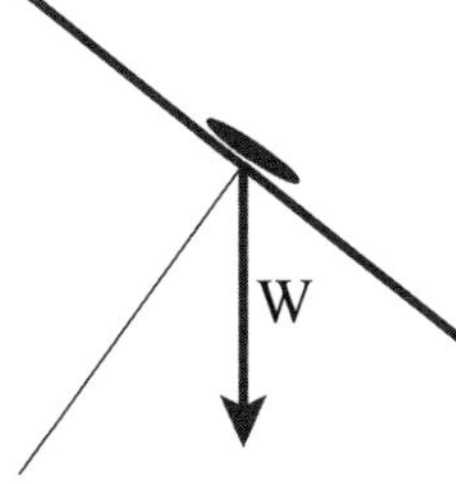

Then you draw the rectangle.

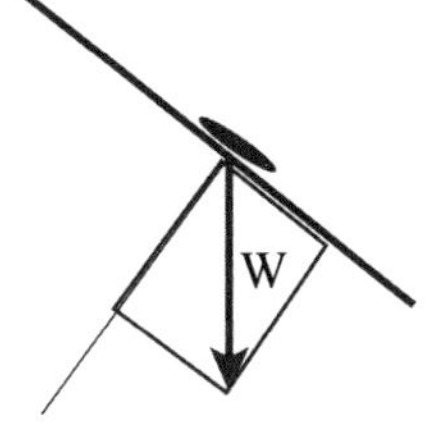

Finally, you draw N and F.

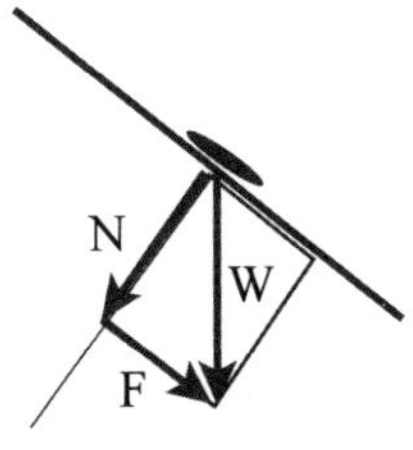

4. $\frac{1}{20}$ psi or 0.05 psi

Pressure is defined as $\frac{\text{weight}}{\text{area}}$

The area is (5)(8) = 40 square inches. Pressure = $\frac{2\text{ lbs.}}{40\text{ in}^2} = \frac{1}{20}$ psi

The Bridge
answers

5. $\frac{1}{4}$

$$F = \mu_k N \qquad \text{We know } F = \frac{1}{2} \text{ and } N = 2$$

$$\frac{1}{2} = \mu_k(2)$$

$$\frac{1}{4} = \mu_k \qquad \text{Dividing both sides by 2}$$

$$\frac{1}{2} \div 2 \Rightarrow \frac{1}{2} \div \frac{2}{1} \Rightarrow \frac{1}{2} \times \frac{1}{2} = \frac{1}{4}$$

6. 6 ft-lb

Work is defined as (Force)(distance) = (2 pounds)(3 feet).

7. Since the distance involved is zero, the work is zero.

8. 3 minutes.

The conversion factor will either be $\frac{60 \text{ pages}}{36 \text{ min}}$ or $\frac{36 \text{ min}}{60 \text{ pages}}$

We want to change 5 pages into minutes, so we want the pages to cancel.

$$\frac{5 \text{ pages}}{1} \times \frac{36 \text{ min}}{60 \text{ pages}} = \frac{5 \cancel{\text{pages}}}{1} \times \frac{36 \text{ min}}{60 \cancel{\text{pages}}} =$$

$$\frac{\overset{1}{\cancel{5}} \text{ pages}}{1} \times \frac{36 \text{ min}}{\underset{12}{\cancel{60}} \text{ pages}} = 3 \text{ minutes}$$

9. $\sqrt{81} = 9$. (See the end of Chapter 3, after the *Your Turn to Play*.)

10. 20 $\qquad 0.6 = 0.03x$

Divide both sides by 0.03 $\qquad \frac{0.6}{0.03} = \frac{0.03x}{0.03}$

Simplify $\qquad 20 = x$

$0.03\overline{)\,0.6}$ becomes $3.\overline{)\,60.}$ with quotient 20.

from p. 139—*fourth try*

1. Inertia is a measure of how much mass an object has. The piece of concrete has the same amount of matter regardless of where it is. The inertia will stay the same.

2. The weight of an object depends on gravity. Since the gravity of the moon is about one-sixth of the gravity on earth, the piece of concrete would weigh about one-sixth as much on the moon.

The Bridge
answers

3. 20 foot-pounds of work was done. Work is defined as force times distance. $W = Fd = (4 \text{ lbs.})(5 \text{ ft.}) = 20$ ft-lb
4. The energy of motion was converted into the energy of height, which is often called potential energy.
5. 0.75 or $\frac{3}{4}$

$$F = \mu_s N \qquad 3 = \mu_s 4 \qquad \frac{3}{4} = \mu_s$$

6. 200 cubic feet

The volume of a box is the length times the width times the height.
$V = lwh = (10')(4')(5') = 200 \text{ ft}^3$

7. 55°

The sum of the angles of any triangle equals 180°.
Since the angles at Jane and Jake add up to 125°, that leaves 55° for the angle at Jack.

8. 48 pounds per square foot

P = density times height = $(40 \text{ lbs./ft}^3)(1.2 \text{ ft}) = 48 \text{ lbs./ft}^2$

9. Two and half hours

Given earnings of \$8/hour and a purchase of \$20, the question is whether to add, subtract, multiply, or divide. Restate the problem using simpler numbers: How long would it take to buy \$6 if you earned \$2/hour? It would take 3 hours. Divide the purchase price by the earnings per hour.

Divide \$20 by \$8/hour.

$$\begin{array}{r} 2\tfrac{1}{2} \\ 8\,\overline{)\,20} \\ \underline{16} \\ 4 \end{array}$$

10. $50x = 25$

Divide both sides by 50 $\quad \frac{50x}{50} = \frac{25}{50}$

Simplify $\quad x = \frac{1}{2}$

from p. 141—*fifth try*

1. 0.126 foot-pounds

Work equals force times distance.

$W = Fd = (0.07 \text{ lbs.})(1.8 \text{ feet}) = 0.126$ ft-lb

2. $\frac{1}{4}$ foot (or 3 inches)

Pressure equals density times the height of the surface above the object. $P = dh$ Given $P = 9 \text{ lbs./ft}^2$ and $d = 36 \text{ lbs./ft}^3$

$P = dh$ becomes $9 = 36h$.

Divide both sides by 36 $\quad \frac{9}{36} = h \qquad$ So $h = \frac{1}{4}$

The Bridge
answers

3. $3\frac{1}{2}$ seconds

It is traveling at 2 inches per second and goes 7 inches. Here are two different ways to solve the problem.

First way: If you don't know whether to add, subtract, multiply, or divide, restate the problem with easier numbers. Suppose it was traveling at 2 inches per second and went 6 inches. That would take 3 seconds. We divided the distance traveled by the rate. So in the original problem we divide 7 inches by 2 inches per second.
$7 \div 2 = \frac{7}{2} = 3\frac{1}{2}$

Second way: Probably the most famous formula from beginning algebra is distance equals rate times time. $d = rt$

The distance, d, is equal to 7 inches. The rate is equal to 2 inches per second.

$d = rt$ becomes $7 = 2t$

Divide both sides by 2 $\frac{7}{2} = t$ So $t = 3\frac{1}{2}$

4. \$88

The conversion factor will either be $\frac{6 \text{ in}^2}{\$8}$ or it will be $\frac{\$8}{6 \text{ in}^2}$

We are starting with 66 in² so we pick the conversion factor so that the square inches cancel.

$$\frac{66 \text{ in}^2}{1} \times \frac{\$8}{6 \text{ in}^2} = \frac{\$(66)(8)}{6} = \$88$$

5. Two-thirds of a pound.

Weight is directly influenced by gravity. In outer space where there is no gravity, objects don't have any weight. Since the moon's gravity is one-sixth of earth's gravity, things will weigh one-sixth as much.

$\frac{1}{6} \times 4$ lbs. $= \frac{4}{6}$ lbs. $= \frac{2}{3}$ lbs.

The Bridge
answers

6.

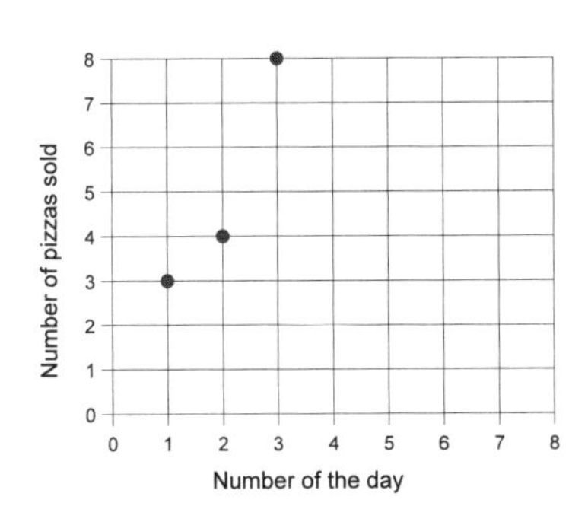

or

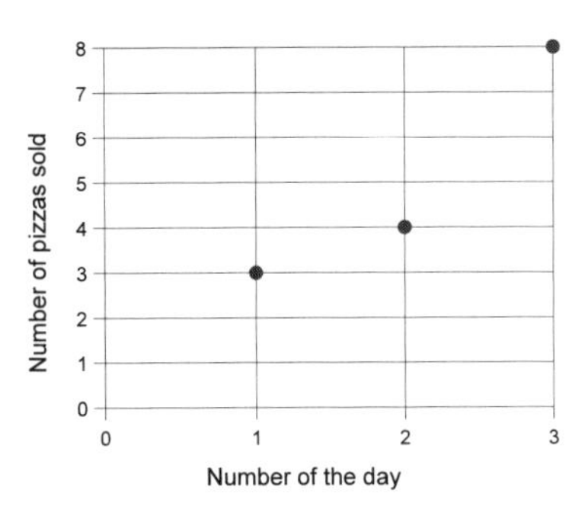

or

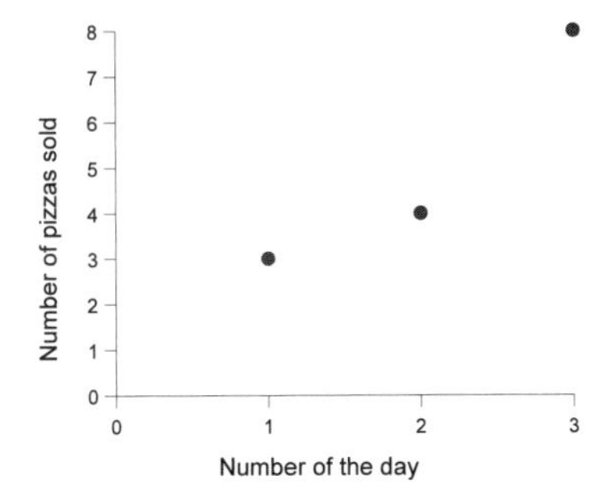

There is more than one way to draw a graph.

7. Discrete. You can sell 7 pizzas or 8 pizzas but not 7.893 pizzas.

8. 12,000 cubic inches

This is Cavalieri's Principle: the volume equals the area of one of the poker chips times the height. $(150 \text{ in}^2)(80 \text{ in.}) = 12{,}000 \text{ in}^3$.

9. 5 pounds per square foot

Pressure is defined as weight divided by area. Thirty pounds spread over an area of 6 square feet will produce a pressure of 5 lbs/ft^2.

10. $\sqrt{9} = 3$ since $3 \times 3 = 9$.

from p. 171—*first try*

1. 3%

360 is ?% of 12,000. We don't know both sides of the *of*, so we divide the number closest to the *of* into the other number.

$$12000 \overline{)\,360.00} = 0.03 \qquad \begin{array}{r} 36000 \end{array} \qquad 0.03 = 3.\% = 3\%$$

2. $\frac{3}{4}$ or 0.75

$F = \mu_s N$ Given F = 600 lbs. and N = 800 lbs.

$600 = \mu_s(800) \qquad \mu_s = \frac{600}{800} = \frac{3}{4} = 0.75$

The Bridge
answers

3. 750 lbs.

$F = \mu_s N$ Given $\mu_s = 0.75$ (from the previous problem) and N = 1000 lbs. (800 lbs. + 200 lbs.).

$F = (0.75)(1000) = 750$ lbs.

4. 0.6 square meters (which can be written as 0.6 m^2)

Area of a rectangle = length times width = $\ell w = (0.8)(0.75) = 0.6$

5. The kinetic energy of the submarine was converted into heat by friction. (If your sentence contained the word *friction*, it was probably correct.)

6. 18 miles per hour (which can be written as 18 mph)

distance equals rate times time.	$d = rt$
Given $d = 3$ miles and $t = \frac{1}{6}$ hours.	$3 = r(\frac{1}{6})$
Multiply both sides by 6	$18 = r\frac{1}{6} \times 6$
Simplify	$18 = r$

7. 620 $lbs./ft^2$

Fluid pressure equals density of the fluid times the height of the surface above the object.

$p = dh$ (Answer to the famous BIG QUESTION)

Given $d = 62$ $lbs./ft^3$ and $h = 10$ ft.

$p = dh = (62)(10) = 620$ $lbs./ft^2$

8. Yes.

The buoyancy is equal to the density times the volume of water displaced. $b = dv$. It only depends on those two things. It doesn't depend on how far under the water the object is. (The pressure on the object does depend on how far under the water it is—but that is an entirely different question.)

9. 0.16 or $\frac{16}{100}$ or $\frac{4}{25}$

Slope is defined was $\frac{\text{Rise}}{\text{Run}}$ which in this case is $\frac{16}{100}$

10. $\sqrt{36} = 6$ since $6 \times 6 = 36$.

The Bridge
answers

from p. 173—*second try*

1. 3 gold coins

The conversion factor will either be $\frac{\$555}{15 \text{ coins}}$ or $\frac{15 \text{ coins}}{\$555}$ since 15 coins = \$555. We start with \$111 and chose the conversion factor that will make the units cancel.

$$\frac{\cancel{\$}111}{1} \times \frac{15 \text{ coins}}{\cancel{\$}555} = \frac{\overset{1}{\cancel{111}}}{1} \times \frac{15 \text{ coins}}{\underset{5}{\cancel{555}}} = 3 \text{ coins}$$

2. 0.18 ft-lb

Work equals force times distance. w = fd To lift a 0.06-pound coin 3 feet into the air requires (0.06)(3) = 0.18 ft-lb of work.

3. 0.01 pounds

One-sixth of 0.06 pounds is 0.01 pounds.

4. 0.18 kilograms.

Kilograms measures the mass of an object—how much matter it has. The amount of matter in a coin stays constant when you move the coin.

5. 20 feet

The pressure of a column of Golden Nectar in a straw is equal to density times height. p = dh When the Golden Nectar is sucked up as high as possible in the straw, this pressure will match the atmospheric pressure on the surface of the liquid. 2000 lbs./ft^2 = dh = (100 lbs./ft^3)h

$2000 = 100h$

$20 = h$ Dividing both sides by 100

6. 30%

There are 6 liters of boysenberry in 20 liters of Golden Nectar.

6 is ?% of 20. When you don't know both sides of the *of*, you divide the number closest to the *of* into the other number.

$$20\overline{)6.0} = 0.3 \qquad \begin{array}{r} 6\,0 \\ \hline \end{array}$$

0.3 = 30.% = 30%

7. 0.02 pounds

Buoyancy is equal to density of the liquid times the volume of the liquid that is displaced. b = dv

The buoyancy experienced by the coin is b = (100 lbs./ft^3)(0.0004 ft^3) = 0.04 lbs.

The weight of the coin is 0.06 lbs. 0.06 – 0.04 = 0.02 pounds.

The Bridge
answers

8. $\frac{1}{3}$ $F = \mu_k N$ $0.02 = \mu_k(0.06)$ $\frac{0.02}{0.06} = \mu_k$

$\frac{0.02}{0.06} = \frac{2}{6}$ (multiplying top and bottom by 100)

9. 1.6 pounds $F = \mu_s N$ $F = (0.4)(4) = 1.6$

10. $x = \frac{1}{8}$

$16x = 2$

Divide both sides by 16 $x = \frac{2}{16} = \frac{1}{8}$

from p. 175—*third try*

1. If the sponge floats, then its density must be less than the density of the bath water. It must be less than 62 pounds per cubic foot.

2. 0.0372 pounds

Buoyancy is equal to the density of the liquid times the volume of the liquid that is displaced. $b = dv = (62)(0.0006) = 0.0372$

(The fact that his toe is 6 inches underwater doesn't matter. If his toe had been 20 inches underwater, the buoyancy would still be 0.0372 pounds.)

3. 0.72 Calories

The conversion factor will either be $\frac{1.2 \text{ Calories}}{5 \text{ hours}}$ or $\frac{5 \text{ hours}}{1.2 \text{ Calories}}$

$$\frac{3 \text{ hours}}{1} \times \frac{1.2 \text{ Calories}}{5 \text{ hours}} = \frac{3 \cancel{\text{hours}}}{1} \times \frac{1.2 \text{ Calories}}{5 \cancel{\text{hours}}}$$

$$= \frac{(3)(1.2) \text{ Calories}}{5}$$

$$= 0.72 \text{ Calories}$$

4. $\mu_s = \frac{3}{4}$ See the third line on page 86.

5. 1% (or 1.0%)

The question is 0.37 is ?% of 37. We don't know both sides of the *of*, so we divide the number closest to the *of* into the other number.

$$37 \overline{)\,0.37} \quad \text{quotient } 0.01; \quad 37$$

$0.01 = 1.\%$

The Bridge
answers

6. 1,620 square inches (or 1,620 in^2)

Area of a rectangle equals length times width. $A = lw$

7. 10.1 minutes

The conversion factor will either be $\frac{2 \text{ gal}}{1 \text{ min}}$ or $\frac{1 \text{ min}}{2 \text{ gal}}$

$$\frac{20.2 \text{ gal}}{1} \times \frac{1 \text{ min}}{2 \text{ gal}} = \frac{20.2 \cancel{\text{gal}}}{1} \times \frac{1 \text{ min}}{2 \cancel{\text{gal}}} = 10.1 \text{ minutes}$$

8. 93 pounds per square foot

Fluid pressure equals density of the fluid times the height of the surface above the object.

$p = dh$ (Answer to the famous BIG QUESTION)

Given $d = 62$ lbs./ft^3 and $h = 1.5$ ft.

$p = dh = (62)(1.5) = 93$ lbs./ft^2

The volume of his toe does not matter in this problem.

9. 40°. The three angles of a triangle always add to 180°. If the first two angles (70° + 70°) add to 140°, then the third angle must be 40° since 70° + 70° + 40° = 180°.

10. $\sqrt{100} = 10$ since $10 \times 10 = 100$.

from p. 177—*fourth try*

1. Inductive reasoning. Jake had gone from observations/trials/ experiments to a conclusion. His conclusion is only probably true.

2. Deductive reasoning. Jane started with three given statements (① You thought it would always be sunny. ② It is raining now. ③ It is not sunny.) and arrived at the conclusion: You were wrong.

With deductive reasoning, if the given statements are true, then the conclusion will be true—not just probably true.

3. $\mu_k = 0.2$

$F = \mu_k N$ $\quad 36 = \mu_k 180$

Dividing both sides by 180 $\quad \frac{36}{180} = \mu_k$

$$180\overline{)36.0} = 0.2$$
$$360$$

4. 0.028 cubic inches

Volume of a box equals length times width times height.

$V = lwh = (0.7)(0.4)(0.1) = 0.028$

5. 37.5% (or 37½%)

15 is ?% of 40 If you don't know both sides of the *of*, you divided the number closest to the *of* into the other number. $\frac{3}{8}$

In *Life of Fred: Decimals and Percents*, you memorized the Nine Conversions:

$\frac{1}{2}$ = 50%

$\frac{1}{3}$ = 33⅓%

$\frac{2}{3}$ = 66⅔%

$\frac{1}{4}$ = 25%

$\frac{3}{4}$ = 75%

$\frac{1}{8}$ = 12½%

$\frac{3}{8}$ = 37½% = 37.5%

$\frac{5}{8}$ = 62½%

$\frac{7}{8}$ = 87½%

6. 31 lbs./ft^2

Fluid pressure equals density of the fluid times the height of the surface above the object.

p = dh (Answer to the famous BIG QUESTION)

Given d = 62 lbs./ft^3 and h = 0.5 ft.

p = dh = (62)(0.5) = 31 lbs./ft^2

7. 0 ft-lb of work

Work is defined as force times distance. w = fd Since he wasn't moving his Electro-Gizmo™, there was no work involved.

8.

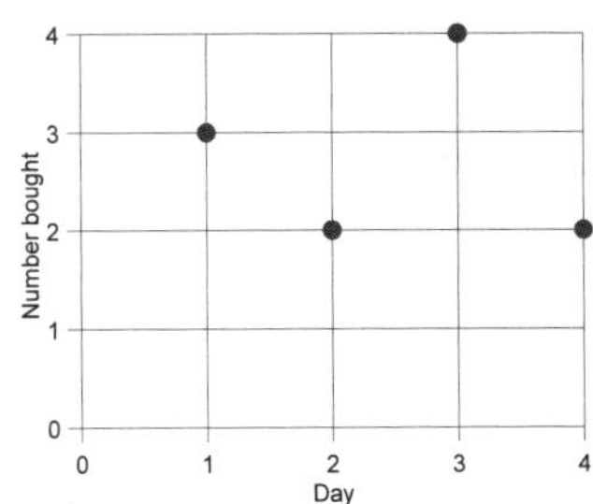

9. The energy of height (also known as gravitational potential energy) was converted into the energy of motion (also known as kinetic energy).

10. $\sqrt{4} = 2$ since $2 \times 2 = 4$

from p. 179—*fifth try*

1. 135 square inches

   Area of a rectangle equals length times width. $A = \ell w = 9 \times 15$

2. The density of the dough must be greater than the density of the oil if the dough sinks rather than floats. The density of the dough is greater than 40 lbs./ft^3.

3. 160 lbs./ft^2 (or 160 pounds per square foot)

   Fluid pressure equals density of the fluid times the height of the surface above the object.

   $p = dh$ (Answer to the famous BIG QUESTION)

   Given $d = 40$ lbs./ft^3 and $h = 4$ ft.

   $p = dh = (40)(4) = 160$ lbs./ft^2

4. 6 ⅔ minutes (or 6 minutes and 40 seconds)

   The conversion factor will either be $\dfrac{4 \text{ minutes}}{60\%}$ or $\dfrac{60\%}{4 \text{ minutes}}$

   $$\frac{100\%}{1} \times \frac{4 \text{ minutes}}{60\%} = \frac{(100)(4) \text{ minutes}}{60} = \frac{400 \text{ minutes}}{60}$$

   $$= \frac{40}{6} \text{ minutes} = \frac{20}{3} \text{ minutes} = 6\tfrac{2}{3} \text{ minutes}$$

5. 6 ft-lb of work

   Work is defined as force times distance. $w = fd = (2)(3)$

6. 5 pounds

   $F = \mu_k N$ Given $F = 2$ and $\mu_k = 0.4$ $2 = (0.4)N$

   Divide both sides by 0.4 $\dfrac{2}{0.4} = N$

   $0.4\overline{)2.0}$ becomes $4.\overline{)20.}$

The Bridge
answers

7. This is inductive reasoning. Inductive reasoning moves from making observations, which is what Stanthony did, to conclusions.

8. 12 ft-lb of work.

Work is defined as force times distance. $w = fd = (6)(2) = 12$
Work can be done vertically (as in this problem) or horizontally as in problem 6.

9. In his best week of sales, he sold 40 **Crispy Pizzas**.

10. $5y = 30$

Divide both sides by 5 $\quad \frac{5y}{5} = \frac{30}{5}$

Simplify both sides $\quad y = 6$

from p. 215—*first try*

1. $27\frac{3}{11}$ ohms

$$\frac{1}{R} = \frac{1}{R_1} + \frac{1}{R_2} = \frac{1}{50} + \frac{1}{60} = \frac{6}{300} + \frac{5}{300} = \frac{11}{300}$$

$R = \frac{300}{11}$ which reduces to $27\frac{3}{11}$

2. 720 lbs./ft^2

Using a conversion factor (see the footnote on page 168),

$$\frac{5 \text{ lbs.}}{in^2} \times \frac{144\ in^2}{1\ ft^2} = \frac{5 \text{ lbs.}}{\cancel{in^2}} \times \frac{144\ \cancel{in^2}}{1\ ft^2} = 720 \text{ lbs./}ft^2$$

3.

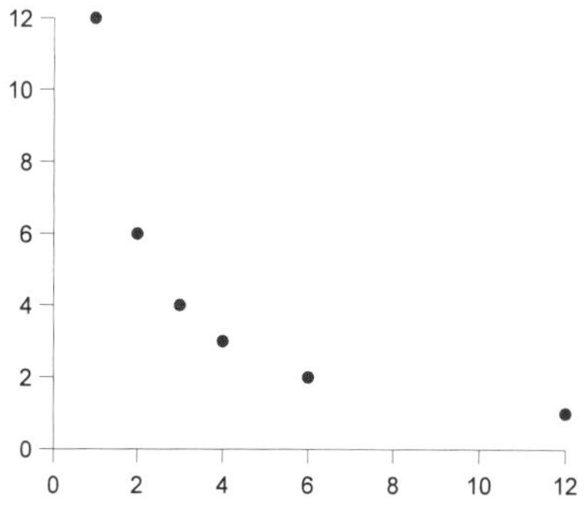

4. $\mu_s = 2/3$

$F = \mu_s N$ $\qquad 3.2 = \mu_s(4.8)$

Divide both sides by 4.8 $\qquad \frac{3.2}{4.8} = \mu_s$

The Bridge answers

$\frac{3.2}{4.8}$ = [multiplying top and bottom by 10] $\frac{32}{48}$ = [Dividing top and bottom by 16] $\frac{2}{3}$

5. We need to know the volume of the radio.

If the density of the radio is greater than the density of water, then the radio will sink. We know the weight of the radio. If we know its volume, then we can find its density.

For example, if the volume of the radio were 0.08 cubic feet, then the density of the radio would be $\frac{4.8 \text{ lbs.}}{0.08 \text{ ft}^3}$ which is 60 lbs./ft^3 and the radio would float. 60 lbs./ft^3 < 62 lbs./ft^3

If the volume of the radio were 0.06 cubic feet, then the density of the radio would be $\frac{4.8 \text{ lbs.}}{0.06 \text{ ft}^3}$ which is 80 lbs./ft^3 and the radio would sink. 80 lbs./ft^3 > 62 lbs./ft^3.

6. 83%

500 is ?% of 600. We don't know both sides of the *of*, so we divide the number closest to the *of* into the other number.

```
      0.833
600) 500.00
     4800
      2000
      1800
       2000
       1800
```

0.833 = 83.3% ≐ 83%

Or it could be done: $\frac{500}{600} = \frac{5}{6}$

and $\frac{5}{6}$ equals 83⅓% (which you may have memorized in *Life of Fred: Decimals and Percents*) which rounds to 83%.

7. The Law of Conservation of Energy can be stated in several different ways: ☞ Energy can't disappear or be created.

☞ The number of foot-pounds of energy in the universe always stays the same.

☞ In any closed system, the amount of energy stays constant.

8. $\mu_s = 0.6$

From Chapters 13, 14, and 15, we learned that $\mu_s = \frac{\text{Rise}}{\text{Run}}$ which in this case is $\frac{4.8}{8}$

9. Hooke's law (Chapter 9)

10. $x = \frac{2}{5}$

from p. 217—*second try*

1. 15 Ω When resistors are in series, just add.
2. $\frac{4}{5}$ amps or 0.8 amps.

First of all, in a series circuit it doesn't matter where you put the ammeter. All of the electrons flow through each of the bulbs.

V = IR becomes $\quad 12 = I(15)$

Dividing both sides by 15 $\quad \frac{12}{15} = I$

3. $\frac{5}{3}$ Ω or 1⅔ Ω

$$\frac{1}{R} = \frac{1}{R_1} + \frac{1}{R_2} + \frac{1}{R_3} = \frac{1}{5} + \frac{1}{5} + \frac{1}{5} = \frac{3}{5}$$

$$R = \frac{5}{3}$$

4. 62.5% or 62½%

The whole cake weighs 80 pounds. 50 is ?% of 80.

Since we don't know both sides of the *of*, we divide the number closest to the *of* into the other number.

```
     0.625
80)50.000
   480
    200
    160
     400
     400
```

0.625 = 62.5% = 62½%

Or it could be done: $\frac{50}{80} = \frac{5}{8}$ and $\frac{5}{8}$ = 62½% (which you may have memorized in *Life of Fred: Decimals and Percents*)

5. 434 pounds of force

$F = \mu_s N = (0.7)(620) = 434$

6. 496 pounds

Buoyancy = (density of the liquid)(volume of the object)
= dv
= (62)(2)

$= 124$ pounds.

$620 - 124 = 496$

7. Deductive reasoning occurs when you start with statements that you accept as true and from them deduce conclusions that must follow from those starting assumptions.

8. $28

Sixteen copies of *Darlene* are 112 letters. (16 × 7 letters = 112)

$$\frac{112 \text{ letters}}{1} \times \frac{\$3}{12 \text{ letters}} = \frac{112 \cancel{\text{letters}}}{1} \times \frac{\$3}{12 \cancel{\text{letters}}}$$

$$= \frac{\$336}{12}$$

$$= \$28$$

9. A discrete variable.

He couldn't put on 32.8908 letters.

10. **5** is the larger numeral.

(8 is the larger number.)

from p. 219—*third try*

1. $9\frac{3}{13}$ ohms

$$\frac{1}{R} = \frac{1}{R_1} + \frac{1}{R_2} + \frac{1}{R_3} = \frac{1}{20} + \frac{1}{30} + \frac{1}{40} = \frac{6}{120} + \frac{4}{120} + \frac{3}{120}$$

$$\frac{1}{R} = \frac{13}{120} \qquad R = \frac{120}{13} = 9\frac{3}{13}$$

2. The resistance of his computer was 55 ohms.

$V = IR$ $\qquad 110 = 2R$

Divide both sides by 2 $\qquad 55 = R$

3. 28 in^3

By Cavalieri's Principle (Chapter 21), the volume is 4 × 7 cubic inches.

4. The second way is to measure the inertia of the object. Determine the force needed to change an object's speed or direction. In outer space where there is little or no gravity, giving a push to an object and seeing how fast it moves is the easiest way to determine its relative mass.

The Bridge
answers

5. 60 minutes

The conversion factor would either be $\frac{8 \text{ minutes}}{6 \text{ poems}}$ or it would be $\frac{6 \text{ poems}}{8 \text{ minutes}}$

We choose the one that cancels the poems, since we want to turn 45 poems into minutes.

$$\frac{45 \text{ poems}}{1} \times \frac{8 \text{ minutes}}{6 \text{ poems}} = \frac{45 \text{ poems}}{1} \times \frac{8 \text{ minutes}}{6 \text{ poems}}$$

$$= \frac{\overset{15}{\cancel{45}} \times 8 \text{ minutes}}{\underset{2}{\cancel{6}}}$$

$$= 60 \text{ minutes}$$

6. $\mu_k = \frac{4}{7}$

$F = \mu_k N$ $\quad 0.04 = \mu_k(0.07)$

Divide both sides by 0.07 $\quad \frac{0.04}{0.07} = \mu_k$

$\frac{0.04}{0.07} =$ [Multiplying top and bottom by 100] $\frac{4.}{7.}$

7. It is not possible. All we know is that the coefficient of static friction is probably greater than $\frac{4}{7}$

8. 4 lbs.

By Hooke's law, F = kx where F is the force applied to the spring and x is the distance the spring is stretched.

F = kx becomes 2.4 = k(6).

Dividing both sides by 6, we have 0.4 = k $\quad 6\overline{)2.4}$ = 0.4

F = kx becomes F = 0.4x.

When Fred stretches the spring 10 inches, F = 0.4x becomes F = (0.4)(10) which is 4.

9. The spring entered the plastic region. (If it had only reached its elastic limit, it would have come back to its original shape.)

10. When you move from a series of experiments, trials, or observations to a conclusion, you are using inductive reasoning.

The Bridge
answers

from p. 221—*fourth try*

1. 63 pounds

$F = \mu_k N = (0.7)(90) = 63.$

2. 270 foot-pounds

$W = Fd = (90)(3) = 270$

3. 80 cubic feet

Cavalieri's Principle: volume = area of one of the poker chips times the height. V = (area of the base)(height). V = (2)(40) = 80

4. 15%

60 is ?% of 400. If we do not know both numbers next to the *of*, we divide the number closest to the *of* into the other number.

$$\begin{array}{r} 0.15 \\ 400\overline{)60.00} \\ \underline{400} \\ 2000 \\ \underline{2000} \end{array} \qquad 0.15 = 15.\% = 15\%$$

5. 20 Ω

$$\frac{1}{R} = \frac{1}{R_1} + \frac{1}{R_2} + \frac{1}{R_3} + \frac{1}{R_4} + \frac{1}{R_5} = \frac{1}{100} + \frac{1}{100} + \frac{1}{100} + \frac{1}{100} + \frac{1}{100}$$

$$\frac{1}{R} = \frac{5}{100} = \frac{1}{20}$$

$$R = 20$$

6. $558,300.

$$\frac{10 \text{ tons}}{1} \times \frac{\$55830}{1 \text{ ton}} = \frac{10 \cancel{\text{tons}}}{1} \times \frac{\$55830}{1 \cancel{\text{ton}}} = 10 \times 55830$$

7. The number of galleries is a discrete variable. You can have 51 galleries or 52 galleries, but you cannot have 51.87 galleries.

8. y = ⅓

$$6 = 18y$$

Divide both sides by 18 $\quad \dfrac{6}{18} = \dfrac{18y}{18}$

Simplify $\quad \dfrac{1}{3} = y$

9. $\sqrt{49} = 7$ since $7 \times 7 = 49$.

10. 248 lbs./ft^2

pressure = density of the liquid times the height of the liquid above the object = dh = (62 lbs./ft^3)(4 ft.) = 248 lbs./ft^2

from p. 223—*fifth try*

1. The number of toppings is not a continuous variable. You can have five toppings or six toppings, but you cannot have 5⅔ toppings.

2. $60

The conversion factor will either be $\frac{60 \text{ bell peppers}}{\$9}$ or it will be $\frac{\$9}{60 \text{ bell peppers}}$

$$\frac{400 \text{ bell peppers}}{1} \times \frac{\$9}{60 \text{ bell peppers}} = \frac{400 \text{ bell peppers}}{1} \times \frac{\$9}{60 \text{ bell peppers}}$$

$$= \frac{\overset{20}{\cancel{400}}}{1} \times \frac{\$9}{\underset{3}{\cancel{60}}}$$

$$= \frac{20}{1} \times \frac{\overset{3}{\cancel{\$9}}}{\underset{1}{\cancel{3}}} = \$60$$

3. 2⅔ Ω

$$\frac{1}{R} = \frac{1}{R_1} + \frac{1}{R_2} + \frac{1}{R_3} = \frac{1}{6} + \frac{1}{8} + \frac{1}{12} = \frac{4}{24} + \frac{3}{24} + \frac{2}{24} = \frac{9}{24} = \frac{3}{8}$$

$R = \frac{8}{3} = 2⅔$

4. 26 Ω When resistors are in series, just add.

5. $I = 0.5$ or $\frac{1}{2}$

$V = IR$	$6 = I(12)$
Divide both sides by 12	$\frac{1}{2} = I$

6. $x = \frac{3}{8}$

Given	$8x = 3$
Divide both sides by 8	$x = \frac{3}{8}$

7. 5 is a larger number than 4.

8. 288 lbs./ft^2

Using a conversion factor of $\frac{144 \text{ in}^2}{1 \text{ ft}^2}$ (see the footnote on page 168),

$$\frac{2 \text{ lbs.}}{1 \text{ in}^2} \times \frac{144 \text{ in}^2}{1 \text{ ft}^2} = \frac{2 \text{ lbs.}}{1 \cancel{\text{in}^2}} \times \frac{144 \cancel{\text{in}^2}}{1 \text{ ft}^2} = \frac{288 \text{ lbs.}}{1 \text{ ft}^2}$$

The Bridge
answers

9. 130 square feet. Area = length times width = ℓw = 5 × 26 = 130

10. $\mu_s = \frac{5}{12}$

From Chapters 13, 14, and 15, we learned that $\mu_s = \frac{\text{Rise}}{\text{Run}}$ which in this case is $\frac{10}{24}$ which reduces to $\frac{5}{12}$

from p. 224—*sixth try*

Yogi Berra once said, "You can see a lot by just looking."*

Did you notice the open switch on the right side of the circuit? No electrons are flowing through the bottom half of the diagram. That will cut our work in half. The diagram now looks like this:

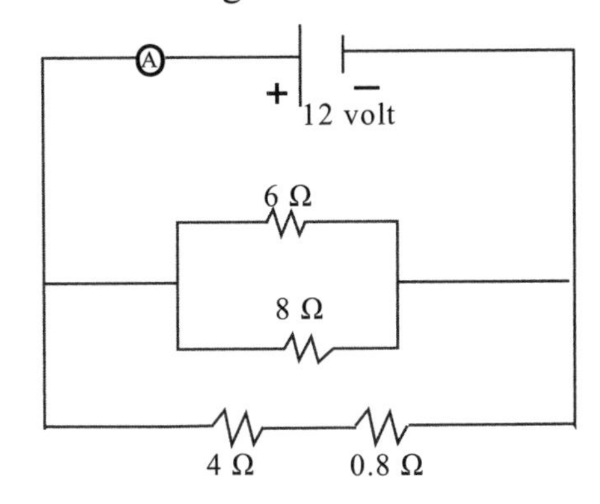

Combining the two resistors on the bottom that are in series we get:

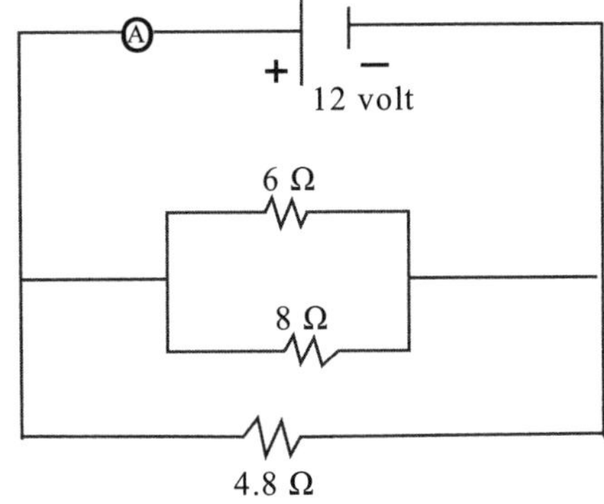

---

* He also said, "You better cut the pizza in four pieces because I'm not hungry enough to eat six."

The Bridge
answers

Combining the 6 Ω and 8 Ω resistors which are in parallel:

$$\frac{1}{R} = \frac{1}{R_1} + \frac{1}{R_2} = \frac{1}{6} + \frac{1}{8} = \frac{4}{24} + \frac{3}{24} = \frac{7}{24}$$

$R = \frac{24}{7}$ and the diagram becomes:

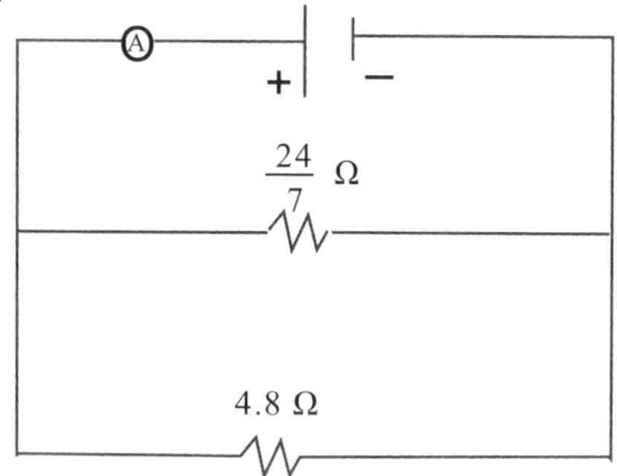

Finally, combine these two resistors which are in parallel:

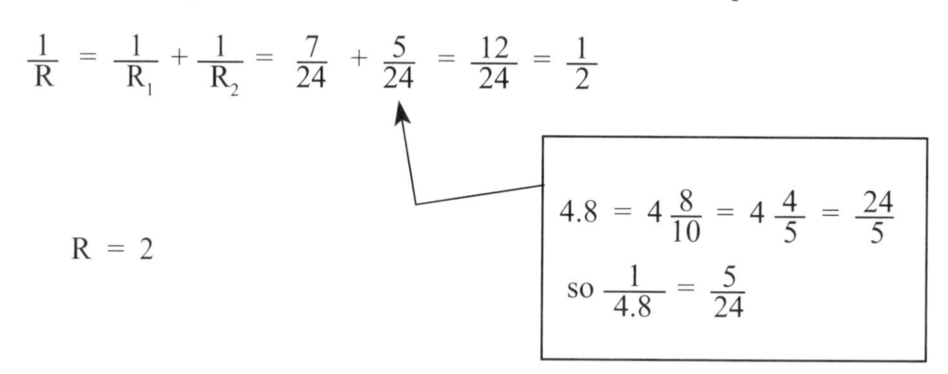

$$\frac{1}{R} = \frac{1}{R_1} + \frac{1}{R_2} = \frac{7}{24} + \frac{5}{24} = \frac{12}{24} = \frac{1}{2}$$

$$4.8 = 4\frac{8}{10} = 4\frac{4}{5} = \frac{24}{5}$$

$$\text{so } \frac{1}{4.8} = \frac{5}{24}$$

$R = 2$

The diagram now reduces to:

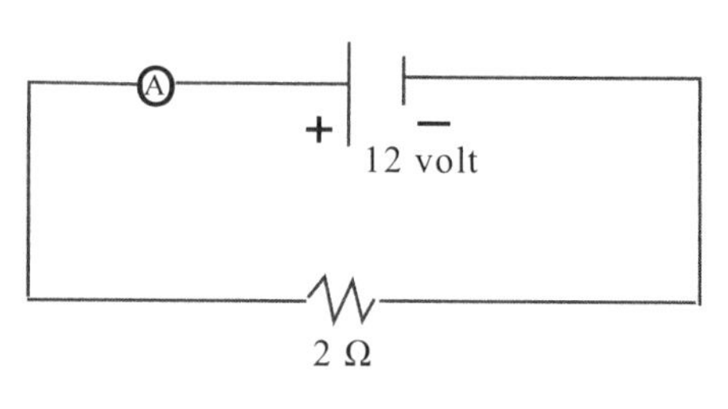

$V = IR \qquad 12 = I(2)$

$6 = I$

The ammeter reads 6 amps.

Index

> greater than. 36, 154
a priori. 44
absolute truth in physics. 48, 49
absolute zero (temperature). 92
acceleration of an object. 58
adding decimal numbers. 186
adding vectors. 82, 84
Alessandro Volta. 188, 190
alpha particles. 231
ammeter. 197, 199, 203
amount of fiction in the Life of Fred books. 43
amperes. 197, 198, 201, 211
angles in any triangle always add up to 180°. 87, 89, 101, 106, 140, 176
animal electricity. 188
Arabic numerals. 36
area of a rectangle. . 16, 38, 39, 41, 42, 68, 69, 72, 74, 104, 136, 171, 179, 223
Arnold Sommerfeld. 239
atomic weight. 228
billion. 128, 229
brain juice (cerebrospinal fluid). . . . 156
breaking point--stretching springs. 53, 67, 74, 220
buoyancy. 147, 148, 152-154, 157, 172, 175, 215, 217
c = the speed of light in a vacuum. 19, 101
Calories. 98, 100, 101
car windshields. 143
Carl David Anderson. 244
Cavalieri's Principle. . . . 131, 132, 137, 141, 147, 219, 222
Celsius temperature scale. 92
Central Park. 72
Christina Rossetti
    "A Royal Princess". 40
    "Advent". 219
    "Goblin Market". 175
    "In the Willow Shade". 220
    "Later Life #16". 70
    "Later Life #21". 137
    "Memento Mori". 104
closed systems. 95, 108, 120
coefficient of friction. 38-42, 72
coefficient of kinetic friction. . . 65, 95, 101, 104, 106, 108, 136, 174, 177, 219, 221
coefficient of static friction. 65, 69, 74, 101, 102, 108, 140, 171, 174, 215, 217
coefficient of static friction found using slope. . . . 75-88, 103, 106, 133, 175, 216, 223
connection between fossils and astronomy. 50
constant of proportionality for friction 26
continuous variables. . . . 19, 23, 39, 41, 42, 68, 69, 73, 74, 102, 107, 108, 136, 141, 218, 222, 223
conversion factors. . . 92, 100-102, 107, 108, 110, 111, 116, 133, 135, 138, 140, 141, 152, 168, 173, 175, 176, 179, 189, 195, 199, 218, 219, 222, 223, 228
converting a mixed number into an improper fraction. 52
cubic inches and cubic feet. 152
cylinder 131
d = rt. 31, 41, 141, 172

Index

degrees in an angle. 231
Democritus. 229, 231, 241, 244
density of liquid water vs. density of ice. 162
density of water. 152, 167
digits in a number. 36, 105
Dirac's equation. 243, 244
discrete variables. . . 19, 38, 40, 69, 73, 102, 107, 108, 136, 141, 218, 222
elastic limit—stretching springs. 52, 53, 67, 74, 93, 220
electrical engineering major. 225
electromagnetic radiation. 238, 240
elliptical constructions. . 184, 185, 193, 208, 228
Energy Cards game. 59-62
Ernest Rutherford. 230, 231, 233
etiquette. 181
exponents. 128, 133
F = kx stretching a spring. 48
F = μN. 18, 29, 30, 35
first coordinate and second coordinate 45
foot-pounds. . . . 90-92, 94, 95, 97, 100, 101, 109, 111
forenoon. 13
friction into heat. 57
friction is independent of speed. 17
friction is independent of the area of contact. 20
Georg Ohm. 205, 206, 215
Giant Chart of Energy. . . . 59, 68, 139, 178
gibbous moon. 27
Gregorian calendar. 112
hard *c* and soft *c*. 26
hard *g* and soft *g*. 26
helium. 238
Henri Becquerel. 230, 231
History of the Atom. 229-244
Hooke's law. 51, 67, 69, 216, 220
How do you know that China exists? 79
hunch—conjecture—theory—law. 48, 49
hung vs. hanged. 40
idiom 13
imperial system. 22
inductive and deductive reasoning. 50, 73, 74, 101, 102, 106, 136, 177, 179, 218, 220
inertia. 117, 118, 139, 219
International Bureau of Weights and Measures. 114
Isaac Newton. 119, 230, 231
J. J. Thomson. 230, 231
James Chadwick. 244
Jeanette MacDonald. 163
joules. 109, 111
Kelvin temperature scale. 92
kilograms. 109-111, 113-115
kilowatt-hours. 97
kinetic friction. 56
kryptonite 23
Law of Conservation of Energy. 94-97, 120, 216
Law of the Conservation of Matter. . . 49
leap seconds. 112
leap years. 112
liter. 111
Louis de Broglie. 240
Luigi Galvani. 187, 188, 190, 191, 193
m, cm, and nm. 236
mass. 110

Index

mass—how to measure it. . . . 113, 114, 117, 136
matter. 95
meter (defined). . . . 21-23, 25, 101, 111
metric system. 52, 109
Metropolitan Museum of Art. 221
million. 128, 229
Mirror Poem. 27
nature hates a vacuum. 163
Nemesis. 50
neutron. 244
neutron star.. 129
newtons. 109
Newton's First Law. 119
Newton's Second Law. 118
Niels Bohr. 233-235
normal force. 79, 80, 82
North Pole and West Pole. 79
numerals. 14, 35, 38, 105, 218
ohmmeter. 207
ohms. 202
Ohm's Law. . . . 206, 207, 211, 214, 219
omega. 202
orbital magnetic quantum number. . 240
orbital quantum number. 240
parallel circuits. 210-212
parallelepiped. 131
Paul Dirac.. 243
Pauli exclusion principle. 241
percent problems. 162, 165, 166, 169-171, 173, 176, 177, 185, 195, 216, 217, 222
period of a pendulum. 22, 24, 72
perpendicular. 79, 81
perpetual motion machines. 96, 99
photon.. 240
photosynthesis. 98
physics experiment. 78
pi is approximately 3.14159265358979 . . . . 111, 153
plastic region—stretching springs. . 53, 67, 74, 93, 220
plotting points.. . . . . 24, 45, 51, 69, 74, 101, 105, 108, 134, 141, 178, 180, 215
positron.. 244
pressure (definition). . . . 123, 126, 136, 137, 142, 144, 145
pressure in a fluid is equal to density times height. . . 132, 133, 135, 140, 141, 144, 146, 167, 169, 172, 177, 179, 222
principal quantum number. . . . 234, 240
proportional 18, 118, 119, 201
proportional limit—stretching springs 52, 67, 74, 93, 220
psi.. 124
quadrillion. 198, 229
question mark at the end of a quotation 193
quintillion. 198
quotient.. 76
QWERTYUIOP. 203
radio waves.. 238
resistance in parallel circuits. . 210-215, 217, 219, 222, 223
resistors. 201
right angle.. . . . . 79, 80, 87, 89
Roman numerals.. 36
Rules of Golf. 152
safe occupations. 77
schematic drawing. 195, 209, 210
Schrödinger's equation.. 242
second of atomic time (defined). . . . 115
second-order partial differential equations. 183

Index

sentence fragments. 236, 241
series circuit. 207, 212
similar triangles. 86, 89
simultaneous events and Einstein. . . . 49
slope of a line. 76, 77, 79, 172
small essays
"A Secret about Being a Good Leader/Boss/Parent". 209
"Areas of Swim Masks". 127
"How Many Atoms Are in That Piece of Bacon?". . . . 228, 229
"Least Dense and Really Dense" 129
"Making Tons of Money". 196
"Math vs. Physics". 14
"The Story of the Meter". 21-23, 25
speaking metaphorically. 235
spectrometer 237
speed of light. 71
spin magnetic quantum number. . . . 240
spring scale. 44
square inches and square feet. 127, 128, 215, 223
square roots. . . 24, 138, 142, 172, 176, 178, 222
static friction. 56, 57
storing energy. 98
stretching springs—the four cases. 52, 53, 67, 74, 220
subshells 239
sucking on a straw. . 160, 161, 163, 167
Super Fred. 34, 43
theorems. 49
Thomas Edison. 219
Thomas Merton. 215
thought experiment. 161
Three-Body Problem. 50
trillion. 229
two kinds of friction. 56
uncertainty principle. 242
V = IR. 35
vectors. 81
voltmeter. 207
volts. 197, 201
volume of a box. 177
volume of a desk lamp. 151
volume of a sphere. 153
wave-particle duality. 183, 240
wavelength 236
wavelength of an electron. 240
wavelengths of various colors. 237
weight density (definition). 130
weight of one oxygen atom. 228
Werner Heisenberg. 242
Wolfgang Pauli. 241
work (as defined in physics). . . . 90, 91, 94, 101, 108, 111, 137, 139, 141, 173, 177-180, 221
X-rays. 238